UNDP/UNCHS/World Bank
Urban Management Programme

Strategic Options for Managing the Urban Environment

18

Toward Environmental Strategies for Cities

Policy Considerations for Urban Environmental Management in Developing Countries

Carl Bartone
Janis Bernstein
Josef Leitmann
Jochen Eigen

Published for the Urban Management Programme by
The World Bank, Washington, D.C.

This document has been prepared under the auspices of the United Nations Development Programme/United Nations Centre for Human Settlements (Habitat)/World Bank–sponsored Urban Management Programme. The findings, interpretations, and conclusions expressed here are those of the authors and do not necessarily represent the views of the United Nations Development Programme, UNCHS, the World Bank, or any of their affiliated organizations.

Deputy Director
Division for Global
and Interregional Programmes
United Nations Development
Programme

Chief
Technical Co-operation
Division
United Nations Centre
for Human Settlements
(Habitat)

Chief
Urban Development Division
Transport, Water, and Urban
Development Department
Environmentally Sustainable
Development
The World Bank

First printing June 1994

The Urban Management Programme (UMP) represents a major approach by the United Nations family of organizations, together with external support agencies (ESAs), to strengthen the contribution that cities and towns in developing countries make toward economic growth, social development, and the alleviation of poverty. The program seeks to develop and promote appropriate policies and tools for municipal finance and administration, land management, infrastructure management, environmental management, and poverty alleviation. Through a capacity building component, the UMP plans to establish an effective partnership with national, regional, and global networks and ESAs in applied research, dissemination of information, and experiences of best practices and promising options.

The findings, interpretations, and conclusions expressed in this paper are entirely those of the author(s) and should not be attributed in any manner to the World Bank, to its affiliated organizations, or to members of its Board of Executive Directors or the countries they represent. The World Bank does not guarantee the accuracy of the data included in this publication and accepts no responsibility whatsoever for any consequence of their use. Some sources cited in this paper may be informal documents that are not readily available. The boundaries, colors, denominations, and other information shown on any map in this volume do not imply on the part of the World Bank Group any judgment on the legal status of any territory or the endorsement or acceptance of such boundaries.

ISSN: 1020-0215

Library of Congress Cataloging-in-Publication Data

Toward environmental strategies for cities : policy considerations for urban environmental management in developing countries / Carl Bartone ... [et al.].
p. cm. — (Urban management programme. Strategic Options for Managing the Urban Environment ; 18)
Includes bibliographical references.
ISBN 0-8213-2827-1
1. Environmental policy—Developing countries. 2. Environmental degradation—Developing countries. 3. Cities and towns—Developing countries. I. Bartone, Carl. II. UNDP/UNCHS/World Bank Urban Management Programme. III. Series: Urban Management Program. Strategic Options for Managing the Urban Environment ; 18.
GE190.D44T69 1994
363.7'009172'4—dc20 94-13583
CIP

CONTENTS

EXECUTIVE SUMMARY 1

I. URBANIZATION, DEVELOPMENT, AND THE ENVIRONMENT 8
- Development and the Environment 8
- Poverty and the Environment 10
- The Brown Agenda 11
- Issues of Spatial Scale 14
- Urban and Regional Ecosystem Considerations 14
- Economic Status and Types of Environmental Problems 16
- Key Actors—Their Roles and Concerns 20
- Cross-Jurisdictional and Cross-Sectoral Issues 22
- Political and Economic Tradeoffs 23

II. ANALYZING THE PROBLEMS AND CAUSES 25
- Understanding the Problems 25
- Factors Aggravating Urban Environmental Degradation 32
- Defining Environmental Impacts and Setting Priorities 36
- Priority Environmental Problems and Policy Linkages 38

III. OPPORTUNITIES TO IMPROVE THE URBAN ENVIRONMENT 41
- Improving the Urban Environment: Key Principles 41
- Mobilizing Public Support and Participation 43
- Improving Governance—Building Institutions and Improving Urban Services 46
- Choosing Appropriate Policy Instruments 59

IV. FORMULATING AN URBAN ENVIRONMENTAL MANAGEMENT STRATEGY 65
- Urban Environmental Planning and Management 65
- Start Up and Assessment 68
- The City-Specific Environmental Management Strategy 72
- Follow-up and Consolidation 79

ANNEX A: CASE STUDIES 86
- Case Study 1: Environmental Consultation in Dar es Salaam 86
- Case Study 2: Rapid Urban Environmental Assessment in the Côte d'Ivoire 89
- Case Study 3: Environmental Management Strategy for Metropolitan Manila 92
- Case Study 4: Water Quality Management in the Delaware Estuary 95
- Case Study 5: Catchment Protection in the São Paulo Metropolitan Region 100
- Case Study 6: Mexico City's Plan for Tackling Air Pollution 103

ANNEX B: ENVIRONMENTAL STRATEGIES WORKING DOCUMENTS 107

REFERENCES 109

BOXES

1.1 Classification of cities by demographic characteristics 9
2.1 Solid waste collection in selected cities 27
2.2 Air pollution in selected urban areas 29
3.1 Creating a constituency for the environment in the Olinda slums 45
3.2 Community provision of low-cost sanitation 47
3.3 Urban environmental management in Tianjin 49
3.4 Integrated planning and environmental improvement in Curitiba, Brazil 54
3.5 Private sector service delivery 57
3.6 Resource recovery by the informal sector 58
4.1 Urban environmental data—outline of questionnaire 70
4.2 Generic outline for urban environmental program 71
4.3 The value of cost-benefit analysis for air pollution control 74
4.4 Cost-effectiveness analysis of Beijing water supply options 78
4.5 External support of pollution control in São Paulo 80
4.6 Multilateral programs supporting urban environmental planning and management 83
4.7 International networks supporting urban environmental planning and management 84
A1.1 The Dar es Salaam declaration 87

TABLES

1.1 Distribution of large cities in developing countries 8
1.2 Protection against environmental hazards to health 18
1.3 Urban population by development level, 1990 18
1.4 Economic-environmental typology of cities 19
2.1 Critical policy links for urban environmental management 39
A2.1 Selection of secondary cities 90

FIGURES

1.1 Spatial scale of urban environmental problems 15
1.2 Environmental indicators at different country income levels 17
1.3 Demand for new cars as a function of GNP 20
2.1 Supplying water to urban areas: current cost and projected future cost 28
4.1 SCP environmental planning and management process 66
A4.1 Costs and benefits associated with water quality objective sets in the Delaware Estuary 98
A6.1 Cost of controlling air pollution from transport in Mexico City 105

FOREWORD

As a direct follow-up to the United Nations Conference on the Environment and Development (UNCED), this paper represents a major effort to extend our understanding of the urban environmental agenda and provide a framework and a process for setting local priorities and formulating environmental management strategies and action plans. It is hoped that cities can use this framework to intitiate their own environmental management programs. The report is intended to give guidance to urban policy analysts, decisionmakers, planners, and managers, as well as to the environmental community and international agencies.

A milestone was achieved at the UNCED Earth Summit (Rio de Janeiro, 1992) when cities were successful in broadening the environmental debate to focus attention on urban priorities. There was broad-based agreement that the developing world's growing urban populations need attention, and that the main concern of those living in cities is the "brown agenda"—involving pollution problems, environmental hazards, and poverty. The Earth Summit also recognized that local authorities and interest groups are best able to take concrete actions on the urban environment. The challenge now is to maintain the momentum built up before and during the Rio conference, and to implement the decisions reached at UNCED. One key recommendation from Rio (that is taken up in this paper) is to develop environmental strategies and action plans for cities. Such approaches will constitute local versions of "Agenda 21," a broad, global framework for environmental improvements and sustainable development.

Prepared for the urban management and environment component of the joint UNDP/UNCHS(Habitat)/World Bank Urban Management Programme (UMP), this paper was written on the basis of specially prepared background and discussion papers, in combination with city case studies and research on a wide range of urban environmental issues. Each paper provides background information on key urban development and environment linkages and suggest elements of an environmental management strategy for cities in the developing world.

The UMP represents a major approach by the United Nations family of organizations, together with external support agencies (ESAs), to strengthen the contribution that cities and towns in developing countries make towards economic growth, social development, and the alleviation of poverty. The program seeks to develop and promote appropriate policies and tools for municipal finance and administration, land management, infrastructure management, environmental management, and poverty alleviation. Through a capacity building component, the UMP plans to establish an effective partnership with national, regional, and global networks and ESAs in applied research, dissemination of information, and experiences of best practices and promising options.

Phase 2 of the UMP (1992-96) is concerned with capacity building at both the country and regional levels and with facilitating national and municipal dialogues on policy and program options. It emphasizes a participatory structure that draws on the strengths of developing country experts and expedites the dissemination of that expertise at the local, national, regional, and global levels.

Through its regional offices in Africa, the Arab States, Asia and the Pacific, and Latin America and the Caribbean, the UMP seeks to strengthen urban management by harnessing the skills and strategies of regional experts, communities, and organizations in the private sector.

Regional coordinators use these networks to address the five program themes in two ways:

- **City and country consultations.** The UMP brings together national and local authorities, private-sector networks, community representatives, and other actors to discuss specific problems within the UMP's subject areas and to propose reasoned solutions. Consultations are held at the request of a country or city, and often provide a forum for discussion of a cross-section of issues.

- **Technical cooperation.** To sustain follow-up to the consultations, the UMP uses its regional networks of expertise to provide technical advice and cooperation.

Through its nucleus team in Nairobi and Washington, D.C., the UMP supports its regional program and networks by synthesizing lessons learned, conducting state-of-the-art research, and supporting dissemination of program related materials.

This paper is intended to contribute to the dialogue that will occur in the city and country consultations being developed in each of the regions. Future work papers will serve as tools to help local practioners apply the strategic framework on the ground.

Mark Hildebrand
Chief
Technical Cooperation Division
United Nations Centre for
Human Settlements (HABITAT)

Louis Y. Pouliquen
Director
Transportation, Water, and
Urban Development Division

ABSTRACT

Rapid urbanization in the developing world, if ignored, can be a threat to health, the environment, and urban productivity. Cities are the engine of economic growth, but the environmental implications of such growth need to be assessed and managed better. The critical and most immediate problems facing developing country cities are the health impacts of urban pollution that derive from inadequate water, sanitation, drainage and solid waste services, poor urban and industrial waste management, and air pollution, especially from particulates. Collectively dubbed the "brown agenda," this set of problems disproportionately affects the urban poor and takes a heavy toll on urban health and productivity.

While the brown agenda is broadly recognized as a universal priority for low-income countries, individual cities may also face many of the "green" issues such as the depletion of water and forest resources, inefficient transport systems and energy use, the degradation of environmentally fragile lands, and the occupation of areas prone to natural hazards. Facing such a multiplicity of problems, it is important for each city to identify specific priorities, policies, and actions needed to address the most immediate issues.

Experience in both industrialized and developing countries demonstrates that an effective approach for confronting urban environmental issues is to formulate a city-specific environmental management strategy and action plan. This paper describes a strategic approach to urban environmental planning and management that is based on participation, building commitment, and choosing effective policy interventions. Five key policy areas are emphasized: (i) mobilizing public support and participation, (ii) choosing policy instruments that will change behavior, relieve conflicts, and encourage cooperative arrangements, (iii) building local institutional capacity, (iv) strengthening urban service delivery, and (v) increasing local knowledge about the urban environment.

The strategic planning process described in this paper has been successfully applied in many cities, and involves several activities:

- *informed consultation* through which environmental issues are clarified, key actors are drawn in, political commitment is achieved, and local priorities are set;

- the formulation of an overall *urban environmental management strategy* that provides long-term environmental goals and phased targets for meeting the goals, and agreement on *issues-oriented strategies and actor-specific action plans* for achieving the goals; and

- *follow-up and consolidation* during which programs and projects are initiated and policy reforms and institutional arrangements are solidified.

Case studies are presented to show how institutional, informational, political, and technical problems related to urban environmental management can be addressed in a strategic manner.

ACKNOWLEDGMENTS

The principal authors of this study are Carl Bartone, Janis Bernstein, and Josef Leitmann of the World Bank's Urban Development Division, and Jochen Eigen of UNCHS (Habitat). The document benefitted from external review by David Satterthwaite and Herman van der Tak. The authors are grateful to Patricia Annez, Jose Carbajo, Michael Cohen, John Courtney, Fitz Ford, Surhid Gautam, John Redwood, K.C. Sivaramakrishnan, Andrew Steer, and David Williams for their thoughtful review of various drafts of the text. Merrell Tuck-Primdahl edited the manuscript, Sriyani Cumine and Laura Lewis provided word processing. Brit Saksvig and Michelle Zook handled the production of the final document.

ABBREVIATIONS

CBO	Community-based organization
EAP	Environmental action plan
EIA	Environmental impact assessment
EMS	Environmental management strategy
ESA	External support agency
GDP	Gross domestic product
GNP	Gross national product
GTZ	German Agency for Technical Cooperation
MEIP	UNDP/World Bank Metropolitan Environmental Improvement Programme
NGO	Non-governmental organization
OECD	Organization for Economic Cooperation and Development
SCP	UNCHS(Habitat) Sustainable Cities Programme
UMP	UNDP/UNCHS (Habitat)/World Bank Urban Management Programme
UNCED	United Nations Conference on Environment and Development
UNCHS	United Nations Centre for Human Settlements (Habitat)
UNDP	United Nations Development Programme
UNEP	United Nations Environment Programme
USAID	U.S. Agency for International Development
WHO	World Health Organization

EXECUTIVE SUMMARY

i. Protecting and improving the urban environment is fast becoming a necessity rather than a luxury. Rapid urbanization in the developing world is threatening health, the environment, and urban productivity. Cities are the powerhouse of economic growth, with 80 percent of GDP growth in developing countries expected to come from cities in this decade, but the environmental implications of such growth need to be assessed and managed better. Municipal authorities and national policymakers would do well to learn lessons from other cities and urban regions that are working to stave off serious problems.

Urban Environmental Problems: The Brown Agenda

ii. The critical and most immediate problems facing developing country cities are the health impacts of urban pollution that derive from inadequate water, sanitation, drainage and solid waste services, poor urban and industrial waste management, and air pollution, especially from particulates. Collectively dubbed the "brown agenda," this set of problems is closely linked to the poverty-environment nexus. Important underlying or related issues typically involve inappropriate land uses, precarious housing, deficient public transportation, and road congestion and accidents.

iii. These pressing problems are also related to what may be considered more properly the "green" and the "social" issues of urban areas, such as the depletion of water and forest resources, the degradation of environmentally fragile lands, the occupation of areas prone to flooding or landslides, overcrowding, degradation or loss of historical and cultural property, noise pollution and other problems. Similarly, the brown emissions of cities resulting from energy use for cooking, heating, industry, and transport contribute significantly to such "global" problems as climate change and acid rain.

iv. While the brown agenda is broadly recognized as a universal priority for low-income countries, individual cities may also face many of the other green and social problems identified above. It will be important for each city to identify its specific priorities and the policies and actions needed to address the most immediate issues.

Poverty, Development, and the Environment

v. Poverty, economic development, and the environment are inextricably linked. This linkage raises issues of equity (such as the willingness to pay for better environmental services and the issue of subsidizing basic urban services for the poor), and of the changing nature of environmental problems at different income levels.

vi. The urban poor are affected disproportionately by brown environmental problems and their plight exacerbates urban environmental crises. The survival of the urban poor is strongly linked to the functioning of urban labor markets and to the markets for land and housing. About a quarter of the world's urban population lives in absolute poverty—and many more live in substandard conditions. In many parts of the developing world, urban poverty has grown faster than rural poverty because of macroeconomic adjustment, inefficiencies in the urban economy, and misallocation of public resources. When subsidies are removed due to budget constraints and fiscal reforms, the poor face higher prices for food, shelter, and essential services. The weakest suffer the most. Among the poor, those most vulnerable to environmental threats include women, children, cottage industry workers, and the elderly. Ways must be found to reduce their vulnerability and risk. Confronted by improperly functioning land markets, the poor often have little choice but to occupy hazardous or polluted areas. This lack of access by lower income families to serviced land, affordable shelter, and basic environmental infrastructure and services has plagued fast-growing developing-country cities for several decades.

vii. The persistent neglect of the basic needs of the poor, together with mounting environmental problems, are taking a heavy toll on urban health and productivity. Today 30 percent of urban dwellers—some 450 million people—lack any form of sanitation. Thus, inadequate sanitation is a major cause of sickness in cities and is a drain on urban economies (because of lost work days due to illness and the costs of treating pollution-related illnesses and cleaning up the mess left behind). Other examples of serious environmental effects on productivity abound. According to one estimate, the costs of pollution problems alone in developing countries exceed 5 percent of GDP. Clearly, improving the situation of the urban poor is an essential pre-condition for reducing urban environmental hazards.

Variability in Urban Environmental Problems

viii. The nature and severity of environmental problems as well as the character of potential intervention strategies in any one city will depend on five factors:

- *The unique natural features of urban areas*—cities tend to take their form from the nature of their site, and their environmental problems are determined in large measure by regional ecosystem characteristics. Ecosystem features vary widely, for example, whether coastal or inland, mountainous or flat, arid or humid, temperate or tropical, or some combination of these features. Each characteristic can be associated with potential problems: for example, topology affects drainage; altitude and atmospheric stability are determinants of urban ventilation and air quality.

- *Population size and rate of growth*—the sheer magnitude and pace of population growth in cities directly affects the spatial concentration of people, industry, commerce, vehicles, energy consumption, water use, waste generation, and other environmental stresses, as well as pressure on the local managerial and operational capacity to respond.

- *The level of income and economic development*—as people rise out of poverty, the nature of their environmental problems change. For example, a low level of economic development is associated with inadequate basic services, indoor air pollution, and land degradation, whereas cities with higher levels of income and economic development suffer more from industrial and energy-related pollution of air and water, and the inability to manage hazardous and toxic wastes.

- *The diverse spatial dimensions of environmental problems in cities*—spatial factors determine who is affected and how, the severity of impact, and the appropriate level of responsibility and decisionmaking needed to solve problems. For example, access to basic environmental services is normally a household, neighborhood, or community issue, while air and water pollution are city-wide or regional issues.

- *The roles of local actors*—the quality of the environment in a city is determined largely by the interaction of numerous public, private, and household actors who have an important effect on environmental problems and their solutions. The managerial and operational capacity and responsiveness of local and sectoral institutions is a critical determinant, as is the level of participation of local stakeholders.

Key Policy Messages for Urban Environmental Management

ix. Experience in both developed and developing countries demonstrates that an effective approach for confronting urban environmental issues is to formulate an urban environmental management strategy and action plan. For large cities and metropolitan regions, this will entail a city-specific design focusing on agreed priority problems. For small towns and intermediate cities a common strategy may cover several cities, and sometimes will address a single issue (for example, a national or regional solid waste management plan). The strategic approach to urban environmental planning and management suggested here is based on participation, building commitment, and choosing effective policy interventions. It responds to the need for local versions of Agenda 21, the broad environmental strategy for sustainable development agreed upon at the United Nations Environment Conference on the Environment and Development in Rio in 1992.

x. The six key policy messages that inform this approach center around: (i) public support and participation, (ii) policy interventions, (iii) service delivery, (iv) institutional capacity, (v) the gap in what is known about the urban environment, and (vi) strategic planning.

xi. *Mobilizing public support and participation:* Urban environments are unlikely to improve without constituencies that demand environmental quality and are willing to pay for it, that help curb vested interests, and that hold local institutions accountable for delivering services. Partnerships for finding, implementing and financing solutions should be built among those affected by environmental problems as well as the proposed solutions, those who control policies and implementation instruments, and those who possess relevant information and expertise. In low-income communities, the participation of community groups and the NGOs that support them, is a powerful instrument for bringing about necessary political commitment and implementing affordable solutions.

xii. *Improving policy interventions—making strategic choices.* In designing interventions, policymakers should avoid excessive reliance on regulation alone and seek short-term instruments and incentives that will change behavior, relieve conflicts, and encourage cooperative arrangements. Environmental policies (or their absence) exert a strong influence on the behavior of both public and private actors and therefore on urban environmental quality. In formulating environmental policies, authorities should consider the severity of each problem, including potential health and economic costs, links to other environmental problems, and cost-effective implementation strategies. Because resources are often limited, policymakers should focus on a few strategic interventions or activities that can then be effectively managed. In São Paulo, for example, water resource management was chosen as the sub-sector where strategic intervention was most urgently needed and where several cross-cutting problems could be addressed at once. Policy tools can include economic and regulatory instruments, property rights, land management instruments, and information and education. The choice of instruments will depend on the nature of the problem. Resource depletion issues, for example, are best dealt with through economic instruments, whereas pollution issues will require a balance of regulatory and economic instruments.

xiii. *Building institutional capacity.* As cities grow and develop, so too must their capacity to manage the urban environment. Policy efforts should focus on upgrading local technical and management capabilities and on establishing cooperative arrangements for dealing with the environmental spillovers that cities generate. In particular, attention should be given to upgrading the operational management. Capacity should also be improved in the areas of strategic planning, municipal finance, housing finance, land management, and poverty alleviation.

xiv. *Strengthening service delivery.* This involves upgrading the management of local environmental infrastructure and services for which cities are directly responsible (water supply, sanitation, drainage, solid waste collection and disposal, and traffic management). Efforts in this area should stress "win-win" situations where environmental benefits go hand-in-hand with economic ones. For example, providing services that people want and requiring users to pay the full cost of environmental services (whether directly through fees and charges or collectively through local taxes) will enhance the efficiency and coverage of service delivery. Private sector participation in service delivery, such as through competitive concessions for waste disposal or other operations, can improve environmental accountability and save money.

xv. *Closing the knowledge gap.* Policies in this area should emphasize routine collection, assessment, use, and dissemination of critical (and currently scarce) data. Informed environmental analysis based on adequate data serves several functions. Used by government officials, such knowledge improves policymaking and the management of urban environmental services. It also informs political consensus-building and institutional strengthening efforts. Both public and private actors involved in managing the urban environment, therefore, should strengthen their ability to routinely collect, assess, use, and disseminate the various types of data required for carrying out their environmental management responsibilities. Improving information and understanding also means expanding curricula on the scientific, technical, and managerial aspects of urban environmental services and land management.

xvi. *Planning strategically.* Improving the urban environment requires management strategies and plans that focus on essential interventions that can be implemented quickly, have a high chance of success, lay the groundwork for solving future environmental problems, and give priority to strengthening incipient urban institutions.

Toward an Urban Environmental Strategy

xvii. Broadly, the environmental planning and management approach described here attempts to blend careful analysis with consensus-building and the participation of a diverse cast of actors. Experience in industrialized and developing countries shows that institutional, informational, political, and technical problems related to urban environmental management need to be addressed in a strategic manner. A planning strategy should involve several activities:

- *informed consultation* during which rapid assessments are conducted and environmental issues are clarified, key actors are drawn in, political commitment is achieved, and priorities are set through town meetings and formal consultative gatherings;

- the formulation of an integrated *urban environmental management strategy* that embodies long-term goals and phased targets for meeting the goals, and agreement on *issues-oriented strategies* (that cut across the concerns of various actors) *and actor-specific action plans* (that cut across the various issues) for achieving the targets, including the identification of least-cost project options, policy reforms, and institutional strengthening; and

- *follow-up and consolidation* during which agreed programs and projects are initiated, policy reforms and institutional arrangements are solidified, the overall process is made routine, and monitoring and evaluation procedures are put in place.

xviii. Several international and regional initiatives are under way to demonstrate, record and disseminate successful cases of urban environmental planning and management. The Urban Management Programme (a joint effort of UNDP, UNCHS (Habitat) and the World Bank) is aimed at strengthening local governance and environmental planning and management through partnerships at the city level. The Urban Management Programme sponsored this paper and has played a key role in helping cities formulate environmental management strategies and in building regional technical cooperation capacity to support these efforts. Several UN specialized agencies have programs to help improve the urban environment, such as the UNCHS Sustainable Cities Programme and the World Health Organization's Healthy Cities Programme. The three programs work closely together to maximize their impact. Important regional initiatives are also underway, such as the Metropolitan Environmental Improvement Programme, supported by the UNDP and executed by the World Bank, that is helping to prepare environmental management strategies and mobilize needed investments for six Asian capitals—Beijing, Bombay, Colombo, Jakarta, Kathmandu, and Metro Manila. Similarly, within the Mediterranean Environmental Action Plan (sponsored by the European Investment Bank and the World Bank), the MEDCITIES Network is carrying out urban environmental audits and helping cities to prepare environmental action plans.

Formulating an Approach

xix. Planning and management efforts should consider the roles of a wide range of public and private interests; the costs of alternative policy and technical solutions; the cross-media and cross-sectoral nature of environmental problems; and the roles and capabilities of public institutions. To work, any urban environmental strategy must reconcile three overriding tensions in environmental management (which are usually, in reality, false dichotomies). The first is **integrated versus sector-specific approaches**. An integrated (that is, cross-media and cross-sectoral) approach is needed in analyzing problems and designing strategies and action plans. At the same time certain agreed actions can only be carried out effectively through designated agencies responsible, for example, for sector-specific improvements in urban service delivery and pollution control. The second false dichotomy is **analysis versus process**—the gap must be closed between the findings and recommendations of careful analysis and the concerns of various constituencies demanding actions that may not necessarily reflect the views of the experts. This tension can be resolved through technical support and increased participation. The third tension involves **decentralized versus centralized approaches**. Here, attention must be paid to the roles and comparative advantages of different levels of institutions. While most problems should be dealt with at the community or municipal levels whenever possible, some environmental spillovers can only be dealt with by higher level institutions. Sometimes, for operational reasons, a broad-based agency is needed, as when a river basin or metropolitan authority handles regional pollution control or waste disposal.

Economic and Political Tradeoffs

xx. In confronting urban environmental problems, efforts to reconcile tensions in environmental management are further complicated by the need to make difficult political and economic tradeoffs. These tradeoffs occur at several levels—between the higher productivity of cities due to economies of scale and agglomeration and the increasing costs of providing environmental infrastructure and services, between different strategies and policy instruments for achieving effective environmental management and political will and economic expediency. Important tradeoffs are also required when making difficult choices about how to allocate scarce resources for solving environmental problems or other needs. Such allocation decisions require broad-based agreement on local priorities.

xxi. Political tradeoffs must not be given short shrift. Politicians are acutely aware that, for all environmental decisions that change the status quo, there will be winners and losers. The losers often will be powerful special interest groups who have access to or are part of the political machinery and have gained from regulatory measures drafted to protect their special interests. Similarly, public investments often benefit those with a political voice. The clearest example is that of upper- and middle-income households that enjoy subsidized water, sanitation, and waste collection services while the poor go without. In making the decision to remove subsidies and reallocate resources, local politicians may risk losing support.

xxii. The challenge is to make the transition from "green" politics that sometimes favor better-off special interests to a genuine form of "environmental justice" that is pro-poor. The authors support the approach that empowers the poor through informed consultation. Indeed, if they are politically organized, the urban poor can create strong interest groups that can effectively lobby for shelter, services, and neighborhood infrastructure.

xxiii. Even when political commitment to environmental protection exists, budget constraints force difficult choices—for example, whether to invest in a safe waste disposal site or education or health services programs, or whether to build new roads in response to growing consumer demand or invest in less-polluting public transport systems. Decisionmaking, therefore, will require a realistic assessment of the urgency, full costs, and likely benefits of alternative interventions. It also will require consideration of cultural and political factors as well as complex interactions among the many influential public and private actors.

To Sum Up

xxiv. This paper proposes ways of linking proven planning techniques into a phased process that relies on informed participation and consultation, beginning with rapid assessment and start-up, proceeding through strategic planning, and resulting in the preparation of action plans that are actor-specific and identify both high priority investments and needed policy and institutional improvements. To succeed, this process must be supported by adequate financial and technical investments, policy reform, and better institutional arrangements for environmental planning and management.

xv. To formulate and implement urban environmental strategies and action plans, cities will need to integrate environmental considerations into existing responsibilities and initiate new environmental actions or programs that address critical urban environmental problems. This will require stronger institutions, better facilities and equipment, and incentives for improved institutional performance. And cities must mobilize financial resources for both capital and recurrent expenses. In all this, cities must focus on strategic interventions—on setting goals that are achievable, even when funding is scarce.

xxvi. Resources for meeting broad environmental quality goals will come in part from national sources, particularly when the environmental benefits spill over local boundaries. Yet, for the most part, local beneficiaries will be expected to pay for improvements in environmental infrastructure and services through user charges and benefit taxes. To encourage efficiency and cost recovery local authorities will have to make tough resource allocation decisions and focus on careful financial planning and management. Financing from external support agencies, while only capable of meeting a very small percentage of needs, can be leveraged effectively by channelling such funding toward meeting the need for basic urban environmental services as well as to support policy reform, institution-building, and research. In financing these improvements, cities should ensure that investments extend environmental services to the urban poor through increased efficiency, affordability, and attention to users' willingness to pay for better services.

I. URBANIZATION, DEVELOPMENT, AND THE ENVIRONMENT

1.1 This paper describes the environmental challenge in developing country cities, discusses why environmental degradation persists in these cities, identifies priority actions and alternative instruments that respond to these factors, and defines a process by which city managers and other actors can formulate strategies for reversing environmental deterioration. Targeted in part at a broad audience of professionals involved in urban environmental improvements, this paper is intended primarily for urban policy makers and managers working with large cities in the developing world. The four chapters are divided into a description of urban environmental problems (chapter 1), an analysis of their causes (chapter 2), a presentation of policies and options for dealing with such problems (chapter 3), and an outline of a proven environmental management process through which solutions can be carried out (chapter 4).

Development and the Environment

1.2 Although a driving force in development, cities are threatened in developing countries by pollution, congestion, and environmental hazards resulting from unprecedented rates of rapid urban growth and industrialization. Reversing the deterioration of the urban environment without slowing economic development will require an environmental policy strategy that takes into account a wide range of actors, difficult political and economic tradeoffs, and a complex set of natural, social, and economic relationships. In particular such a strategy must consider the nexus of poverty and environmental degradation.

1.3 Urgent action is needed, because, as urban populations grow, so do environmental problems. Urban populations and the number of large cities are increasing steadily in developing countries, a trend that carries serious environmental implications. Cities are currently absorbing two-thirds of the total population increase throughout the developing world. By the year 2000, twenty-one cities will have more than 10 million inhabitants—seventeen of those mega-cities will be in developing countries (World Bank 1992). By the year 2030, developing country cities are expected to grow by 160 percent.

Table 1.1. Distribution of large cities in developing countries

City size class (population)	No. of cities		Urban population (millions)		Share of urban population (%)	
	1990	2000	1990	2000	1990	2000
> 4 million	33	50	280	455	20.2	23.2
2 to 4 million	43	72	120	197	8.6	10.0
1 to 2 million	111	157	152	215	11.0	11.0
0.5 to 1 million	207	243	144	173	10.3	8.8
TOTAL	394	522	696	1,040	50.1	53.0

Source: United Nations (1985)

Box 1.1. Classification of cities by demographic characteristics

Following are the four principal city size classes that occur throughout the developing world, distinguished by population shares, growth rates, jurisdictional complexity, and administrative capacity:

- **Large metropolitan areas:** The number of large urban agglomerations with populations greater than 2 million will grow worldwide from 112 in 1990 to 172 by the end of the century. The share of the urban population living in them is increasing steadily and is expected to grow 4 percent annually in this decade. Much of this growth will be concentrated in the urban peripheries, or the result of the graduation of large cities into this cohort. These metropolises are increasingly concentrated in the developing world. Large metropolitan areas encompass several municipal jurisdictions.
- **Large cities:** Cities with populations ranging from 500,000-2 million are more numerous and faster-growing than large metropolitan areas. In the developing countries, most cities now in this category will become large metropolitan areas within one to two decades only to be replaced by a new cohort of intermediate cities. There are some 330 developing country cities in this class, expected to exceed 400 by the year 2000 (see Table 1.1 for the distribution of large cities). About one fifth of urban population is located in this size class, but this share is decreasing as cities migrate to the metropolitan cohort.
- **Intermediate cities:** About 1,400 cities have populations of 100,000-500,000. The number of such cities is growing but their share of urban population is diminishing. About 14 percent of the world's urban population worldwide is located in intermediate cities.
- **Small cities and towns:** There are tens of thousands of urban centers with populations of less than 100,000 and about 36 percent of the world's urban population lives in such towns, but this share is steadily decreasing. Often they are administratively weak. Some, however, are important administrative or agricultural centers experiencing rapid growth.

Two other urban formations are increasingly recognized in developing countries. They include *extended metropolitan regions*, such as the Bangkok and São Paulo metropolitan region, and *extended industrial regions*, such as the Upper Silesian Industrial Region around Katowice, Poland.

Source: United Nations (1985) and National Institute of Urban Affairs (1988)

1.4 The distribution of large cities in developing countries is shown in Table 1.1. Although recent data indicate that the rate of mega-city growth is decreasing, the numbers of new urban dwellers continues to swell (for example, Mexico City's population rises by over 500,000 people every year). The pace of growth of small and intermediate-size cities, meanwhile, shows few signs of slowing, although there is often significant diversity even across cities in the same country. In Box 1.1, a classification of cities by demographic characteristics describes the types of cities in the developing world. This paper focuses primarily on large metropolitan areas, large cities, extended metropolitan regions, and extended industrial regions where half of the developing world's urban population presently lives.

1.5 Cities are both engines of growth and sources of concentrated environmental problems. Urban growth and economic development in cities are driven by the higher productivity of urban households and enterprises vis-a-vis rural homes and businesses, which translates into higher wages and increased job opportunities for urban dwellers. Growth in productivity is also linked closely to national macroeconomic performance, as documented in the World Bank policy paper, *Urban Policy and Economic Development: An Agenda for the 1990s* (1991), and the UNDP strategy paper, *Cities, People and Poverty: Urban Development for the 1990s* (1991). As cities grow, productive activities tend to concentrate in urban centers, where 60 percent of gross domestic product (GDP) is generated by about one-third of the total population. Between now and the year 2000, some 80 percent of GDP growth in the developing world is expected to originate in cities and towns, with higher productivity in cities than in the countryside and higher productivity in large cities than in small towns. Increasing productivity, however, has not eliminated the massive problems of poverty and environmental degradation in cities.

Poverty and the Environment

1.6 As the physical environment in and around cities deteriorates, the urban poor are hit the hardest. The World Bank's *World Development Report 1990* estimates that about a quarter of the total urban population lives in absolute poverty—and many more live in substandard conditions. In many parts of the developing world, urban poverty has grown faster than rural poverty because of macroeconomic adjustment, inefficiencies in the urban economy, and misallocation of public resources. When subsidies are removed due to budget constraints and fiscal reforms, the poor face higher prices for food, shelter and essential services. Confronted by improperly functioning land markets, the poor often have little choice but to occupy hazardous or polluted areas. In Mexico City, the highest concentrations of dust particles in the air are found in the low-income sections of the city (Hardoy and Satterthwaite 1989). Moreover, some studies have found that when public services are not accessible, the peri-urban poor pay private vendors ten to twenty times more for water than those living in areas served by public piped water supplies.

1.7 Foremost among the environmental concerns of the urban poor are health problems resulting from a substandard living environment that does not offer protection from human excreta and other wastes, indoor air pollution, or natural hazards. Intra-urban studies confirm that mortality and morbidity from gastroenteric and respiratory infections (including tuberculosis) and malnutrition are significantly higher for the urban poor than for other urban residents. So, too, are the resulting costs of health care and productivity losses. Among the urban poor, children, women, adolescents, cottage industry workers, and the elderly are particularly vulnerable. In many societies women do most of the domestic work. This may mean staying indoors, tending to cook stoves or fires in poorly ventilated homes, or tending to children sick with malaria, dengue, diarrheas, or other water-borne diseases.

1.8 Poverty contributes to urban congestion and environmental degradation in several ways. one is migration of the rural poor to cities in search of a living. Because the poor lack the financial resources to compete for serviced land and adequate housing, they often occupy illegal settlements on hazard-prone or ecologically fragile lands. In many instances, the location of low-income populations on hazard-prone land has resulted in substantial losses. In Rio de Janeiro, where nearly two-thirds of all *favela* (slum) dwellers occupy the steep slopes surrounding the city, mudslides often claim hundreds of lives and leave thousands homeless during the annual rainy season. In 1988, flooding and landslides due to unusually heavy rains exacted a heavy toll of about 289 dead, 734 injured, and 18,560 homeless. The floods and landslides cost an estimated $935 million. The physical losses severely disrupted Rio's economic activity, particularly in the northern part of the metropolitan region, and left the predominantly low-income population with limited access to schools, health facilities, and basic sanitation (Munasinghe, Menezes, and Preece 1991).

1.9 In a setting of rapid urbanization the challenge of urban environmental management is to safeguard the health, productivity, and quality of life of city dwellers that result from their interactions with the physical (built) and natural environments that surround them, as well as from the changes in those environments induced by human activities. In this context, the most important environmental problems facing cities relate to:

- access to basic environmental infrastructure and services;
- pollution from urban wastes and emissions;
- resource losses such as ground water depletion and land degradation; and
- natural and man-made environmental hazards.

The Brown Agenda

1.10 Collectively dubbed the "brown agenda," this set of problems (involving pollution, environmental hazards and poverty) is linked to development, income growth, and concern for green issues, including intergenerational concerns about global warming and natural resource depletion. On the development-income side, problems of urban waste and pollution are inextricably linked to poverty, productivity, and broad macroeconomic performance. Problems that cross over to "green" or "social" issues typically involve inappropriate land use, precarious housing, deficient public transportation, and road congestion. These are frequently manifested in overcrowding, noise pollution, the degradation of envionmentally fragile lands, the occupation of areas prone to floods or landslides, and the degradation or loss of historical and cultural property. Further, solving the brown issues in cities has crucial implications for resolving many natural resource or green issues that extend beyond urban boundaries. Solving a single priority problem, as with São Paulo's effort to launch a catchment area protection program in the São Paulo Metropolitan Region, can be the launching point for an issue-specific environmental management strategy and for a subsequent long-term urban environmental management strategy (see Case Study 5 in the Annex). Likewise, tackling urban air pollution and designing more energy efficient cities can benefit more than one generation (by reducing carbon dioxide emissions and thus slowing global warming).

1.11 Manifest in cities throughout the developing world, brown problems inflict a high price. For example, just 70 percent of urban dwellers are served by some form of sanitation, with only about 40 percent connected to sewers (World Health Organization 1992). Where there is sewerage, more than 90 percent of the wastewater is discharged without treatment. The costs of these and other pollution problems alone in some developing countries are estimated to exceed 5 percent of GDP (Pearce and Warford 1991). The following are only a few examples of city-specific impacts:

- In Bangkok, excessive exposure to lead causes 200,000-500,000 cases of hypertension, resulting in 400 deaths a year. Rough estimates suggest that children with lead poisoning lose an average of four or more IQ points by the age of seven, with long-term implications for their productivity as adults (World Bank 1992 and U.S. Agency for International Development and U.S. Environmental Protection Agency 1990).

- In Mexico City, annual health costs from air pollution are estimated to exceed $1.5 billion. Abnormally high levels of suspended particulates have caused an average of 2.4 lost work days per person each year and 6,400 deaths every year; lead exposure may contribute to as much as 20 percent of the incidence of hypertension in adults and 29 percent of all children have unhealthy lead levels in their blood (Eskeland 1992; Schteingart 1989; and Margulis 1992).

- In Jakarta, health costs associated with selected air pollutants (lead, suspended particulate matter, and nitrogen dioxide) are estimated to be $220 million a year. This includes the costs of avoidable deaths, restricted activity days, outpatient visits, hospital admissions, respiratory illness, hypertension cases, heart attack cases, asthma attacks, and loss of intelligence in children (World Bank 1993).

- In the peri-urban areas of Peru, the cholera epidemic of 1991, which was due to inadequate sanitation and poor hygiene, caused over 320,000 cases, 2,600 deaths, and an estimated $1 billion in losses from reduced agricultural and fisheries exports and tourism (World Bank 1992).

- In metro Manila, the potential productivity impact of air pollution is approaching an estimated $20 million a year; waterborne contamination is responsible for a potential yearly productivity loss of almost $100 million (World Bank 1993).

1.12 For most developing country cities, the brown agenda has two poles, each of which demands attention. At one extreme, there are the traditional environmental health problems associated with the lack of adequate shelter and services. At the other extreme are the environmental health consequences of rapid industrialization without adequate hazardous waste management, air pollution control, occupational health services, industrial accident prevention, and other protective programs. In contrast to industrialized cities, therefore, cities in developing countries are faced with advanced pollution issues before controls over traditional pollution sources have been put in place and before strong institutions are developed. Each city has a unique mix of these problems that changes as development progresses.

1.13 Coping with the brown agenda in urban areas in developing countries requires consideration of each city's unique set of environmental management issues. Of prime importance are the mutually reinforcing effects of poverty and environmental degradation. Other unique features include the pace and intensity of urbanization, the cross-media and spatial implications of environmental degradation, urban land use and environmental inter-relationships, and the wide range of public and private actors involved in causing as well as solving environmental problems.

Dynamics of the urbanization process—pace and intensity

1.14 The magnitude of urban population growth in developing countries is a direct indicator of the degree of spatial concentration of people, industries, commerce, vehicles, energy consumption, water use, waste generation, and other environmental stresses. Some problems are more likely to grow with city size, as in Bangkok, Jakarta, and São Paulo, where annual car and truck usage per resident is expected to increase geometrically with the size of the urban area. This is a daunting prospect, given that these huge cities are adding hundreds of thousands of new inhabitants a year. Likewise, many large and intermediate cities are confronting enormous environmental problems. The Cameroonian cities of Douala and Yaoundé for example, are grappling with industrial emissions as well as human and household wastes.

1.15 The intensity and scale of urban growth as well as the rapid concentration of urban wastes and emissions hampers pollution prevention and control efforts. (Scale here is the product [per unit area] of population density times per capita waste generation [or resource use]). Economic development exacerbates such problems because the quantity of urban wastes generated per capita also tends to increase steadily with increased per capita income. The scale of the problem exceeds the capacity of local government to collect, treat, and dispose of municipal sewage and solid wastes and to control dangerous wastes and emissions; and it exceeds the capacity of nature to assimilate all of these wastes. An analogous scale problem exists on the input side as a result of the concentrated resource consumption in urban areas. Urban demand for food, water, minerals, fuelwood, fossil fuels, and other resources often affects distant populations, forests, and watersheds.

Urban land use and the environment

1.16 Urban land use decisions are critical determinants of environmental quality. Distortions in urban land markets and ineffective land management policies and practices in developing countries have resulted in:

- air pollution;
- congestion and accidents;
- degradation of environmentally fragile lands (for example, wetlands and coastal resources);
- occupation of hazard-prone areas (for example, steep slopes, flood plains, and vacant land adjacent to polluting industries or waste disposal sites); and
- the loss of cultural resources, open space, and prime agricultural land.

1.17 By failing to consider land use planning as an integral part of a transport strategy, many developing country cities have allowed transport activity, particularly automobile traffic, to shape their growth and development (Lowe 1992). Excessive amounts of land dedicated to traffic circulation leads to problems with drainage, run-off, and flooding, as well as traffic congestion, pollution, accidents, and noise. In Warsaw, for example, the master plan envisaged expanding the city to a distance of more than 70 kilometers to protect residents from exposure to automotive emissions. The right of way between buildings on main thoroughfares was increased to 180 meters. The implementation of this plan, however, substantially increased air pollution from automobiles (Bertaud 1992). Conversely, inadequate road space can result in congestion, accidents, and an increase in exposure to air pollution. In Bangkok, for example, each car is expected to spend an average equivalent of forty-four days each year stalled in ever-increasing traffic jams and the city currently loses about one-third of its potential gross city product due to congestion-induced travel delays—this could rise to a loss of 60 percent of gross city product if no actions are taken (Japan International Cooperation Agency 1990).

1.18 Density and spatial patterns of development also have important implications for various environmental outcomes. For example, high-density development can achieve economies of scale in infrastructure provision, but, if not well-planned and serviced, can impose high costs associated with congestion (for example, the rapid spread of communicable disease due to crowding or more accidental injuries). Lower density development outside of the central city means reduced congestion in residential areas, but higher costs for infrastructure provision, and in the absence of adequate public transport, higher levels of air pollution from automobile traffic. The concentration of industry in a few locations poses serious environmental consequences. In Bangkok, Lima, Manila, Mexico City, and São Paulo, for example, industrial pollution, including the impacts of poorly managed hazardous wastes, imposes serious health impacts in densely populated areas.

1.19 Well-intentioned land use controls can protect populations and sensitive resources or they can exacerbate environmental degradation and the vulnerability of low-income populations to hazards. For example, setting aside large amounts of land for recreation, community facilities, and rights-of-way within subdivisions can result in reduced supply and high urban land prices. In India, the Madhya Pradesh land use regulations resulted in residential standards that were affordable to only the wealthiest 20 percent of households.

1.20 Similarly, enforcing development restrictions on hazard-prone land may substantially reduce the supply of urban land and therefore increase the cost of available land for housing. Unless the government or private sector can provide alternative affordable housing sites or allow for increased densities, the increase in land prices will push the poor to outlying areas where they will be further from job opportunities and urban services or lead to overcrowding of accessible inner city areas. By contrast, if the development restrictions are not enforced, the poor will settle on the hazard-prone areas because such land is often the only remaining available space close to jobs and services.

1.21 Any environmental management strategy should attempt to quantify and account for the capital investment and operating costs associated with varying density and spatial patterns, the costs and benefits of resource protection, and the costs of informal land occupation trends. In assessing alternative solutions, these costs will have to be quantified and then allocated between land users and government. In many cities, however, line agencies lack the skills to make such assessments and are unable to communicate the tradeoffs that might come into play to decisionmakers. This normally results in a continuation of environmental problems.

Issues of Spatial Scale

1.22 Figure 1.1 illustrates a number of characteristic problems related to spatial scale and to the provision, or lack thereof, of key urban infrastructure and services. Considering such dimensions helps pinpoint several important issues for environmental management:

- Health impacts are greater and more immediate at the household or community level and tend to diminish in intensity as the spatial scale increases.

- Immediate equity issues arise in relation to the provision of basic services at the household, community, or city scale; long-term concerns about equity (such as the impact of global warming and resource depletion on future generations) are likely to emerge at the regional or global level.

- Levels of responsibility and decisionmaking should correspond to the scale of impact.

Urban and Regional Ecosystem Considerations

1.23 A city's regional ecosystem is often a critical determinant of the severity of environmental conditions within and around urban areas. The extent of regional variations make it difficult to devise a simple typology applicable to all environmental problems encountered in cities. Therefore, natural ecosystem effects should be carefully evaluated. Major regional ecosystem types often associated with critical environmental conditions include:

Figure 1.1. Spatial scale of urban environmental problems

SPATIAL SCALE	HOUSEHOLD/ WORKPLACE	COMMUNITY	METROPOLITAN AREA	REGION	CONTINENT/ PLANET
Key infrastructure and services	shelter water storage onsite sanitation garbage storage stove ventilation	piped water sewerage garbage collection drainage streets/lanes	industrial parks roads interceptors treatment plants outfalls landfills	highways water sources power plants	
Characteristic problems	substandard housing lack of water no sanitation disease vectors indoor air pollution	excreta laden waters/oils trash dumping flooding noise/stress natural disasters	traffic congestion accidents ambient air pollution toxic dumps	water pollution ecological areas lost	acid rain global warming ozone layer

Source: Bartone (1990)

- *Coastal regions.* Many cities are located along coastal areas or estuarine systems. Of the world's largest 100 cities, thirty-six are in coastal regions. Urban development contributes to port and marine pollution which contributes to eutrophication, fish kills, shellfish contamination and beach pollution, loss of wetlands and wildlife habitat, and shoreline erosion.

- *River basins and riverine areas.* Most non-coastal cities are located along rivers and must confront problems of upstream and downstream water quality and quantity. The city of Bogota, for example, presents severe downstream environmental threats to the namesake river that runs though it. Upstream pollution and damming of the Guarapiranga watershed presents critical pollution problems to São Paulo, since residents rely on the Guarapiranga reservoir for water supply. Lakes are also heavily impacted, as with lake Amatitlan in Guatemala and Beira lake in Colombo. Typical problems arising in such water bodies include eutrophication provoked by domestic and industrial waste.

- *Arid regions.* Many cities are also located in arid or semiarid zones. In Latin America, for example, 60 percent of the urban population live in arid and semi-arid areas with only 5 percent of the region's water resources. In these conditions, there are frequent droughts and flash floods. Urban demands for fuelwood and charcoal result in deforestation; intense competition for limited water arises among cities, industries and agriculture; and greater susceptibility to water pollution can significantly increase the cost of available fresh water resources.

- *Humid tropical regions.* Cities in the humid tropics are particularly vulnerable to routine flooding, which often requires expensive sanitation solutions. Heat and humidity aggravate the outbreak of communicable diseases and pests. Tropical coastal areas are also vulnerable to severe storms.

- *Cold regions.* Large cities with cold winter seasons often encounter urban air pollution problems due to energy use for domestic heating and industry. This is particularly true where low-quality coal is used, as in northern China and Eastern Europe.

- *Mountainous regions.* In mountainous areas, urban development (particularly informal housing) often extends to steep hillsides, leading to erosion and slope instability. Flash floods can also be a problem. In addition, altitude and temperature affect such natural processes as the operational efficiency of internal combustion engines, the formation of intermediate products (like ozone) from emissions, and natural waste assimilation rates.

1.24 These urban ecological types can occur in various combinations and are critical factors in determining which environmental strategy will work best in a particular city. In determining wastewater disposal strategies, for example, sewage reclamation and reuse may be considered as an alternative for an arid coastal city, while a long submarine outfall may be more appropriate for a humid coastal city.

1.25 Other geologic or topographic conditions may be hazardous, for example, major geological fault systems subject to frequent seismic activity or mountainous areas with slope instability during rainy seasons. A topographic example is the occurrence of critical air pollution episodes resulting from restricted ventilation conditions in such high, mountainous urban areas as Mexico City.

Economic Status and Types of Environmental Problems

1.26 An assessment of equity issues and income status reveals important correlations between economic status (low-income, lower-middle income, upper-middle income, and high income) and types of environmental problems. Inadequate water supply, sanitation, and solid waste, and high levels of indoor air pollution and land degradation are associated with a low-level of economic development, while higher levels of economic development are associated with industrial and energy-related pollution. Described below is a typology based on these correlations that can help ensure a better fit between the design of environmental and other urban policies and the priority needs of cities.

Relationship between income and pollution pressures

1.27 The variability of environmental indicators at different country income levels reveals important trends. Figure 1.2 illustrates how urban environmental problems may worsen or improve with income growth. Carbon dioxide emissions and municipal wastes per capita increase as incomes rise. On the plus side, more people gain access to safe water and adequate sanitation. Urban concentrations of particulate matter are relatively high at the low-income level, rise slightly at the low- to middle-income level, then drop markedly as per capita income surpasses $800. Urban

Figure 1.2. Environmental indicators at different country income levels

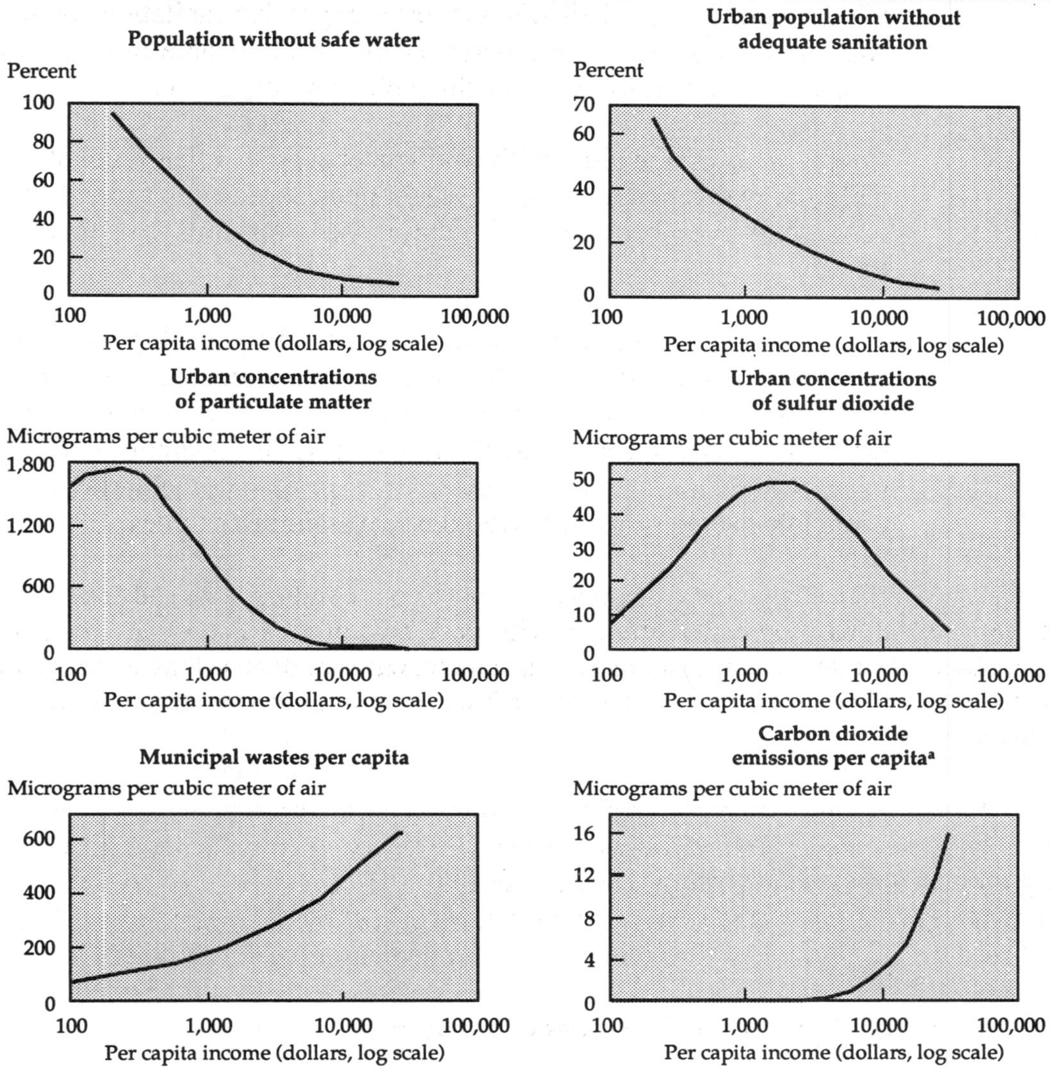

Note: Estimates are based on cross-country regression analysis of data from the 1980s.
a. Emissions are from fossil fuels.

Source: World Bank *World Development Report 1992*

income levels, and gradually decrease after that. Knowing about such trends can help in the formulation of long-range environmental action plans.

Levels of institutional development and income

1.28 The institutional capacity to respond to environmental problems is determined in part by a city's economic development. In developing countries institutions in charge of administering fiscal reforms and promulgating and enforcing environmental policies are generally weak, while institutions in high-income countries have been developing their environmental pollution control policies and financing mechanisms over the past 100 years and now have highly developed capabilities. As cities develop, therefore, the capacity to deal with urban environmental problems should expand in line with the complexity of the problems to be tackled.

Table 1.2. Protection against environmental hazards to health
(Survey of 168 countries by level of economic development)

	No. of countries			
Country status	High-income countries	Middle-income countries	Low-income countries	Total
Most requirements met	27	11	4	42 (25%)
Some requirements met	0	34	14	48 (29%)
Few requirements met	0	15	63	78 (46%)
TOTALS	27 (16%)	60 (36%)	81 (48%)	168 (100%)

Source: Adapted from Shaefer (1991)

1.29 The results of a recent World Health Organization survey (Table 1.2) shows that national and local pollution control capabilities in the middle- and low- income countries need strengthening. A worldwide survey reveals that four out of every five people live in conditions where pollution control capacity is deficient. Generally, the shortfalls are greatest in developing countries and the needs are most urgent where rapid industrialization is under way.

Urbanization pressures and income

1.30 Table 1.3 presents an estimate of urban population by development level in 1990. Clearly, as countries advance, they urbanize and cities burgeon. With the urban population of low-income countries already more than 1 billion, explosive urban growth in the poorest countries seems likely.

Table 1.3. Urban population by development level, 1990

Country status	Urban population (millions)	Proportion urban (% of total)	Population in cities > 1 million (% of total)
High-income countries	628	77	29
Upper-middle-income countries	325	71	30
Lower-middle-income countries	327	52	21
Low-income countries	1,162	38	9
TOTALS	2,443	46	16

Source: World Bank (1992)

Economic-environmental typology of cities

1.31 The level of economic development is a significant determinant of environmental conditions in any city. Drawing on available descriptions of urban environmental problems and impacts, the following presents a general characterization of the different problems and severity of impacts most often encountered in cities of four different income strata—low-income, lower-middle income, upper-middle income, and high-income. Due to the scarcity of GDP city-level data, these four income strata are based on 1990 national GDP per capita data from the *World Development Report, 1992* and are as follows:

Table 1.4. Economic-environmental typology of cities

Urban environmental problems	Lower-income countries (<$650/cap)	Lower-middle-income countries ($650-2,500/cap)	Upper-middle-income countries ($2,500-6,500/cap)	Upper-income countries (>$6,500/cap)
Access to basic services				
• Water supply and sanitation	Low coverage and poor quality, especially for urban poor	Low access for urban poor	Generally acceptable water supply, reasonable sewerage	Good; concern with trace substances
• Drainage	Low coverage; frequent flooding	Inadequate; frequent flooding	Reasonable	Good
• Solid waste collection	Low coverage, especially for urban poor	Inadequate	Reasonable	Good
Pollution				
• Water pollution	Problems from inadequate santiation and raw domestic sewage	Severe problems from untreated municipal discharges	Severe problems from poorly treated municipal and industrial discharges	High levels of treatment; concern with amenity values and toxic substances
• Air pollution	Severe problems in some cities using soft coal; indoor exposure for poor	Severe problems in many cities from soft coal use and/or vehicle emissions	Severe problems in many cities from soft coal use and/or vehicle emissions	Problems in some cities from vehicle emissions; health priority
• Solid waste disposal	Open dumping, mixed wastes	Mostly uncontrolled landfills, mixed wastes	Semi-controlled landfills	Controlled landfills, incineration, resource recovery
• Hazardous waste management	Non-existent capacity	Severe problems, little capacity	Severe problems, growing capacity	Moving from remediation to prevention
Resource losses				
• Land management	Uncontrolled land development and use; pressure from squatter settlements	Ineffective land use controls	Some environmental zoning practiced	Environmental zoning commonplace
Environmental hazards				
• Natural and man-made hazards	Recurrent disasters with severe damage and loss of life	Recurrent disasters with damage and loss of life	High risk from industrial disasters	Good emergency response capacity

1.32 Pollution problems change as technologies advance and as economies develop. In the case of water pollution, for example, the low-income countries are still coping with rudimentary problems of organic waste loads while the most developed countries—which have already developed technologies for dealing with organic wastes, basic industrial wastewater discharges, and nutrient waste loads—are concerned with the effects of low concentrations of toxic substances on water quality.

1.33 Similarly, urban automotive air pollution can be expected to increase with the level of economic development, simply because of the steady increase in motorization and urban automobile ownership. Figure 1.3 shows the growth in automobile ownership with rising incomes. Most of this growth in automobile ownership in developing countries is occurring in the large urban areas—for example, in Peru and Thailand more than 80 percent of motor vehicles are registered in the capital city.

Key Actors—Their Roles and Concerns

1.34 The nature of urban environmental problems is determined in large part by the interaction of numerous public, private, non-governmental, community, and household actors. Some of the most influential actors and their respective limitations (or strengths) include:

- *Environmental protection agencies* responsible for setting environmental regulations and standards and for monitoring and enforcement. Too often such agencies are relatively weak bodies with only coordination functions.

Figure 1.3. Demand for new cars as a function of GNP
(Relationship of GNP to car ownership)

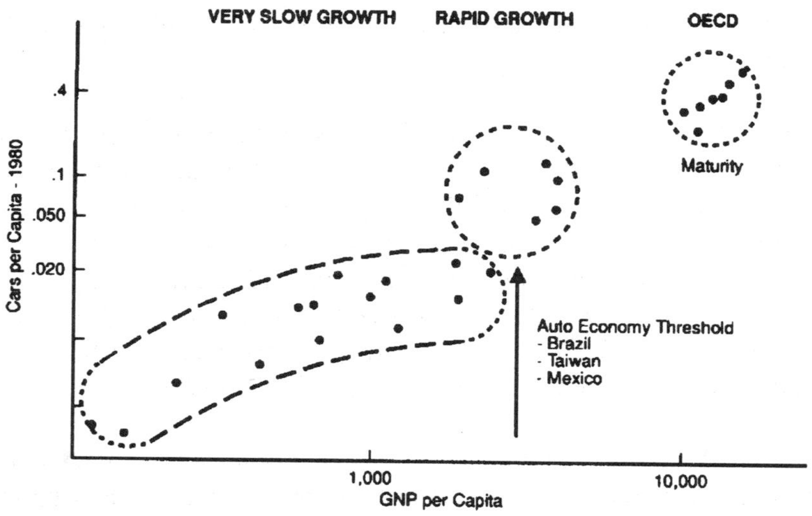

Source: Karmokolias (1990)

- *Planning agencies* whose staff are often unfamiliar with environmental analysis and information and how to apply it to local development planning.

- *Politicians*, particularly at the local level, who are usually preoccupied with maintaining their power and influence and take a short-term view when allocating resources.

- *Sector agencies (public and parastatal)* that tend to have little experience in cross-sectoral collaboration, and give scant consideration to the interrelationships between projects. They will often promote infrastructure and development projects that lack basic environmental considerations.

- *Concerned residents and community-based organizations (CBOs)* whose members and leaders are acutely aware of the impacts of environmental problems at the household and neighborhood level. Typically this group has few opportunities to participate in the preparation of urban infrastructure or industrial projects that affect them.

- *Non-governmental organizations (NGOs)* that can be effective agents for building local awareness, for mobilizing community action, and for voicing local concerns. Often, however, environmental NGOs are focused exclusively on green and global issues and pay insufficient attention to brown issues.

- *Private and informal sector enterprises* that generally are concerned about the constraints placed on their businesses by environmental regulations and the costs such rules incur, particularly when enforcement is lax or inconsistent. Members of the business community seldom wish to be perceived as environmental villains and should be co-opted into the search for solutions.

- *The news media* who voice concern for the environment and in reporting on those affected by environmental conditions. This role can be negative, however, if the tendency is to sensationalize environmental topics rather than focus on real local priorities.

- *The scientific and engineering community*, who play a pioneering role in shaping the environmental agenda. This group often sets its agenda for environmental research and monitoring on the basis of its own scientific interests without giving due thought to the needs of vulnerable populations. Because valuable information on the environment often is published in scientific journals in technical language, the scientific community should ensure that data are communicated in a less scholarly form to wider audiences, including environmental policymakers, planners, and managers, and the media.

- *External support agencies* are a key source of funding for large environmental projects and are important donors of technical assistance and other support. In some instances they may give high priority to the environment and poverty reduction, but many have only recently recognized the critical importance of urban areas to overall national development.

1.34 Analyzing the concerns of each actor and the cultural constraints, social context, and political dynamics in which environmental degradation persists helps reveal the causes of degradation in a particular city. Knowing about these actors will also help in evaluating the full costs of impacts and in estimating the likely benefits of measures to prevent or mitigate damages and improve environmental conditions. Such an understanding of tradeoffs also makes it easier to develop the overall institutional framework for urban environmental management that will suit local needs.

Cross-Jurisdictional and Cross-Sectoral Issues

1.35 The causes and effects of urban environmental problems spread across jurisdictions and across sectors. Excessive vehicular pollution, for example, may stretch into suburban or peri-urban areas. Although they may share the same pollution problems, the satellite communities living in the metropolitan region may not be required to submit to the same emissions testing or car registration requirements as municipal residents. Negotiating cross-jurisdictional solutions involves tough political choices, and (often) awkward alliances of convenience (such as between a regional and municipal government agreeing on landfill location and rates for its use). Urban environmental problems often spread across sectors, which requires coordinated responses from a diverse cast of characters. For example, leachate from a poorly maintained land fill may be leaking into aquifers or a nearby river that supply drinking water to the city. In this way, what began as a sector specific waste disposal problem becomes a cross-sectoral issue with cross-media effects.

1.36 ***Jurisdictional complexity***. Although the diffuse effects of urban pollution require cross-sectoral and cross-jurisdictional coordination, attempts to deal with environmental issues in urban areas are frequently piecemeal and compartmentalized. These problems will grow in scope as cities expand. Large metropolitan areas confront the additional difficulty of spreading across multiple municipal jurisdictions—often as many as thirty. Similar problems will emerge for cities with populations surpassing 500,000 as they expand rapidly and add additional municipal jurisdictions.

1.37 Jurisdictional complexity in urban areas results in part from the multiplicity of actors with overlapping, uncoordinated, or poorly defined responsibilities in environmental management. These actors include national and regional sector agencies, state or provincial governments, and competing local governments. For example, municipal authorities tend to give attention to the environmental concerns of local communities, such as the demand for waste collection services, and ignore air pollution and other responsibilities that spill over into adjacent municipalities or cities located downwind or downstream.

1.38 Jurisdictional conflicts—both across levels of government and between municipalities in metropolitan areas—often impede effective action or give rise to crucial gaps in prevention and control efforts. The example of cross-jurisdictional relations in Brazil is instructive. There, efforts in the 1970s to create metropolitan coordination bodies failed because mayors and governors disagreed about what they interpreted as a partial restriction of their jurisdictional powers. While Curitiba (a Brazilian metropolitan region encompassing fourteen municipalities) has had some success in managing urban services that cross municipal borders, its coordination body, the Coordination of Curitiba Metropolitan Region, has met with limited success at the broader metropolitan level (Rabinovitch and Leitmann 1993). In other instances, the level of authority does not match the spatial scale of impact.

1.39 In assigning environmental responsibilities, the lines between different levels of government are often blurred. Local government may have an important role in regulating pollution from industry (particularly small-scale manufacturing) and keeping hazardous industrial wastes out of municipal sewers and landfills. At the same time, however, municipalities are major sources of pollution and their emissions should be regulated by regional or national environmental authorities. For example, World Bank studies found that in Mexico City, 60 percent of water pollution from organic wastes derives from municipal discharges and only 40 percent from industrial discharges; in Curitiba, (population 2.4 million) only 17 percent of the total organic discharge is from industry.

1.40 ***Cross-media and spatial scale issues.*** Managing the urban environment requires both policymakers and managers to consider the cross-media and spatial scale effects of urban pollution. Any plans to improve one environmental medium (air, water, or land), therefore, should consider the potential effects of that intervention on other media. For example, sewage treatment plants may clean up the flow of wastewater, but produce large quantities of sludge that must be disposed of safely on land. Similarly, emission standards that require the installation of scrubbers will create additional burdens for land disposal of fly ash. In light of cross-media effects, relevant jurisdictions and institutions will have to be carefully coordinated so as to ensure comprehensive management. Failure to do so will lead to both cross-media pollution problems as well as such wasteful, ineffective policies as investments in surface drainage without parallel improvements in solid waste collection and disposal.

Political and Economic Tradeoffs

1.41 Meeting the environmental challenges of urbanization requires making difficult political and economic tradeoffs. These tradeoffs occur at several levels—between the higher productivity of cities due to economies of scale and agglomeration and the increasing costs of providing environmental infrastructure and services; between different strategies and policy instruments for achieving effective environmental management, and between political motives and economic expediency.

1.42 The most typical tradeoffs are made when economic and environmental objectives conflict. Often environmental benefits accrue mainly to downstream, or future users. For example, the payoffs from more stringent industrial emissions standards are manifested in ways that are difficult to monitor immediately (such as improved human health) and require significant sacrifices from users and substantial resources from the government for enforcement and monitoring. By contrast, some innovative solutions involve "win-win" situations in which the goals of environmental protection and of economic efficiency, equity, and growth are complementary. Examples include improvements in the provision of water supply and sanitation services, public transport systems and traffic management, and the introduction of energy efficient technologies in industries and homes.

1.43 Even in win-win situations, a balanced response to the complex issues of rapid urbanization, productivity, poverty, and environmental degradation involves extremely difficult political and economic tradeoffs. For example, local authorities may be reluctant to enforce pollution control regulations, apply the polluter pays principle, or insist on industrial safety programs for fear of driving industry and jobs to other cities.

1.44 Politicians are acutely aware that, for all environmental decisions that change the status quo, there will be winners and losers. The losers often will be powerful special interest groups who have access to or are part of the political machinery and have gained from regulatory measures drafted to protect their special interests. Similarly, public investments often benefit those with political voice. The clearest example is that of upper- and middle-income households that enjoy subsidized water, sanitation, and waste collection services while the poor go without. In making the decision to remove subsidies and reallocate resources, local politicians may fear losing support. By contrast, if they are politically organized, the peri-urban poor also represent a strong interest group that can effectively demand shelter, services, and neighborhood infrastructure. As the poor exert these demands, however, sustained environmental improvements are not guaranteed. Local governments often will give priority to "ribbon-cutting" investments and skimp on meeting the recurrent costs of maintaining local infrastructure or the investments needed to control pollution.

1.45 Even when political commitment to environmental protection exists, budget constraints force difficult choices—for example, whether to invest in safe waste disposal site or education or health services programs, or whether to build new roads in response to growing consumer demand or invest in less-polluting public transport systems. Decisionmaking, therefore, will require a realistic assessment of the urgency, full costs, and likely benefits of alternative interventions. It also will require consideration of cultural and political factors as well as complex interactions among the numerous influential public and private actors.

1.46 To make choices about tradeoffs and priority problems in the face of complex externalities and scarce resources, urban managers must set clear priorities based on the severity of environmental impacts, especially on human health. In Mexico City, because resources for environmental clean up were scarce, urban managers, donors, and other interested actors had to choose between tackling air pollution or water pollution. Both are major problems in the metropolitan area. However, because of great public awareness and concern, the political ramifications of repeatedly being the city with the foulest air in the world and the extremely high costs of air pollution (due to ill health, lost work days, and lost tourism revenues), air pollution emerged as the priority problem. The steps for setting such priorities is the focus of the remaining chapters.

II. ANALYZING THE PROBLEMS AND CAUSES

2.1 The most critical urban environmental concerns in developing countries include problems related to access to basic environmental infrastructure and services, pollution from urban wastes and emissions, loss or destruction of natural and cultural resources, and exposure of urban populations to natural and man-made hazards. These problems are caused in large part by lack of public and political awareness, inadequate governance, inefficient and inadequate economic and regulatory policies, and insufficient knowledge and information. The dynamics of such problems and causes are outlined in this chapter.

2.2 While most cities in developing countries share some of these common problems (and causes), the specific environmental concerns of each urban area are different, depending on their economic status, the prevalence of urban poverty and unequal access to urban services, and other factors. To solve city-specific problems, a set of environmental priorities must be established based on local needs and resources. Problems should be ranked by their effects on human health, productivity, amenity, and ecological values. This should go hand-in-hand with an assessment of the underlying causes (governmental, institutional, economic, and political) of such problems. This chapter thus sets out the first step in defining and ranking environmental impacts. A matrix setting out common problems, causes, and policy implications is set out at the end of this chapter. Opportunities for improving the urban environment are described in detail in chapter three and a process for carrying out environmental action plans and strategies is outlined in chapter four.

Understanding the Problems

2.3 Although the extent of urban environmental problems differs among cities, concern over them is shared by most developing country cities. A review of the extent of urban environmental problems in developing country cities and their causes highlights some prevalent concerns. Specific examples of such problems at the local, regional, and international level illustrate the breadth and depth of environmental degradation.

Access to basic environmental infrastructure and services

2.4 ***Sanitation and sewerage.*** In most low-income cities, the pollutant of primary concern is human excreta (Hardoy and Satterthwaite 1991). The World Health Organization reports that 3.2 million children under the age of five die each year in the developing world from diarrheal diseases, largely as a result of poor sanitation, contaminated drinking water, and associated problems of food hygiene. Infections and parasitic diseases linked to water are the third leading cause of productive years lost to morbidity and mortality in the developing world (World Bank 1993). Diarrheal death rates are typically about 60 percent lower among children living in households with adequate water and sanitation facilities than those in households without such facilities (World Bank 1992). Although about 70 percent of the urban population has access to some form of sanitation, only about 40 percent are connected to sewers (World Health Organization 1992). Where sewerage connections do exist and some treatment is provided, operational difficulties are common and desired effluent quality often is not achieved. In Cairo, for example, 70 percent of the city is connected to the public sewerage system, but only 15 percent of the wastewater collected is fully treated. Of the remainder,

25 percent is partially treated and 60 percent is carried raw via open canals to a lake and then to the sea (Grenon and Batisse 1989). Most domestic sewage is deposited directly on land or into surface water, posing severe public health risks, particularly to children.

2.5 To compound the problem of inadequate domestic wastewater disposal, industrial facilities discharge untreated effluents directly into surface water or on land, causing pollution of both surface and ground water supplies. In the Pakistani cities of Karachi and Lahore, for example, daily industrial discharges into a local stream include about 35 tons of suspended solids, 376 tons of dissolved solids, 2 tons of ammonia, and 1.5 tons of arsenic oxide. At Kala Shah Kaku (near Lahore), industries and manufacturing plants discharge toxic effluents (including mercury) into the Deg Nallah river, making its water unsuitable for irrigation and livestock consumption (World Bank Europe Middle East and North Africa Region 1991). Where industries discharge their untreated water into municipal sewers, the result is damage to the sewer system, often leading to collapse and to interference with municipal treatement systems (where they exist). As industrialization intensifies, conditions can be expected to worsen unless the necessary controls are in place.

2.6 ***Municipal solid waste.*** A persistent problem for local government is inadequate collection and disposal of household garbage (Box 2.1). Although most municipal governments spend between 20-50 percent of their available operating budgets for solid waste services, typically only half of the urban households benefit from collection services. Most wastes that are collected end up in open dumps or drainage systems, threatening both surface water and ground water quality and causing flooding, which provides a breeding ground for disease-carrying pests. Open-air burning of wastes, spontaneous combustion in landfills, and incineration plants that lack effective treatment for gas emissions can cause air pollution. Lack of the most basic solid waste services in crowded, low-income neighborhoods is a major contributor to the high morbidity and mortality among the urban poor.

2.7 ***Storm water.*** In many cities, sewage and sullage from domestic, commercial, and industrial sources is flushed away by drainage systems; the urban runoff is then channeled into surface waters. For many poor people whose houses are built on land unsuitable for residential development, the most dramatic health effect of inadequate drainage is death due to drowning in floods or being buried beneath landslides or collapsing houses. Other health effects include those associated with unsafe sanitation; standing water resulting from poorly drained wastewater and runoff creates ideal conditions for outbreaks of mosquito-borne diseases. In addition to health effects, frequent flooding causes property damage, road congestion, disruption to public services, and lost employment.

2.8 ***Urban transport.*** The inefficiency or lack of urban transport services and infrastructure is a major impediment to economic growth and urban productivity in developing country cities. Increasing motorization, poorly operating public transport services, inadequate road maintenance, insufficient bikeways and walkways, poor traffic management, and lack of enforcement and education contribute factors to congestion, road accidents, and air pollution. In many cities, traffic congestion leads to lost work and leisure time, increased fuel consumption and emissions, and high accident rates. Indeed, road safety is a major concern. The costs of road accidents in developing countries, two-thirds of which occur in urban areas, is as high as 1-2 percent of GDP, taking into account high fatality and injury rates and property damage (WHO 1989).

Box 2.1. Solid waste collection in selected cities

Dar es Salaam: Some two thirds of all solid wastes from both residential areas and from commercial enterprises remains uncollected.

Guatemala City: Of the some 1,100 tons of garbage generated every day, only some 750 tons are collected by private and municipal companies; the rest is thrown into clandestine garbage dumps or left to rot in the raines that surround the city.

Jakarta: About 80 percent of the solid waste is collected and transported to open dump sites. Considerable amounts of waste are buried, burned, thrown on open lands, or dumped into drains, canals, and rivers where it clogs drainage channels and causes extensive flooding during the rainy season.

Katowice: An estimated 70 percent of the residents in the communities of the Voivodeship benefit from refuse collection, with coverage ranging from 40-100 percent in the cities and 0-22 percent in other communities. Throughout the Voivodeship, there are forty-five dumps operated by municipalities, none of which employ sanitary landfill practices.

Port-au-Prince: Only 37 percent of the solid wastes generated is collected. Inhabitants living close to rivers throw their waste directly into the river. In other areas, waste is disposed of through open air burning close to housing. The fumes from incomplete combustion of the waste cause significant air pollution.

São Paulo: Almost 95 percent of the households in the São Paulo Metropolitan Area reported that their waste was collected. The private sector collects and disposes a total of 16,600 tons of solid wastes per day, leaving 3,600 tons to be collected and disposed of informally. Much of this waste is industrial, 20 percent of which is considered hazardous.

Sources: Bogota: Guatemala City: Di Pace et al. (1992), cited in Hardoy and Satterthwaite (1992) Jakarta: Clarke et al. (1991); Katowice: Borkiewicz et al. (1991); Port-Au-Prince: Benoit and Pamphile (1991); Sao Paulo: Leitmann (1991).

Pollution from urban wastes and emissions

2.9 ***Water pollution.*** Pollution of surface waters (that is rivers, lakes, and coastal and marine waters) can result in health problems from eating shellfish or from direct contact with water, and lost revenues as a result of depleted fisheries and a drop off in tourism. And, for many cities, available water resources are being depleted and degraded, thus raising the cost of water use. Figure 2.1 shows that the unit cost of water in several cities would more than double, and in some cases, more than triple under a new water development project. In Lima, for example, upstream pollution has increased treatment costs by about 30 percent. In Shanghai, water intakes have been moved upstream more than 40 kilometers at a cost of about $300 million (World Bank 1992). In many instances, vital ground water resources are compromised by saltwater intrusion or by contamination from toxic or organic waste that seeps into the supply.

2.10 ***Energy use.*** Energy is a key input for urban development, but the supply and demand of virtually every type of fuel generates varying degrees of environmental externalities. Because cities use fuels intensively, energy-related environmental problems are spatially concentrated but often spill over to rural areas. On the supply side, the extraction and conversion of energy resources for urban use can harm the environment in many ways, some of which may occur outside the city (for example, deforestation from fuelwood harvesting or disruption of watersheds and evictions of communities to make way for hydropower projects). On the demand side, human health is affected

Figure 2.1. Supplying water to urban areas: current cost and projected future cost

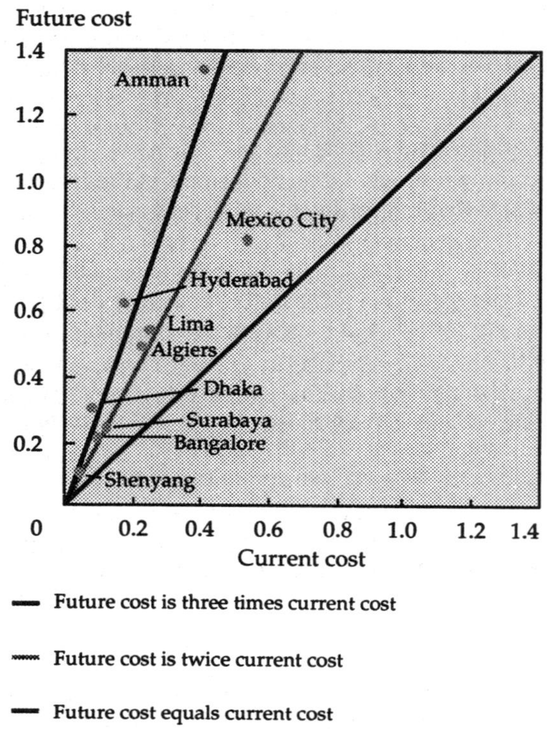

Note: Cost excludes treatment and distribution. Current cost" refers to cost at the time data were gathered. "Future cost" is a projection of cost under a new water development project.

Source: World Bank *World Development Report 1992*

largely because of fuel combustion and its resulting emissions. Secondary impacts include the generation of heat that can raise temperatures in cities by 5-10 degrees centigrade. The nature of these impacts is dynamic: as economic development occurs, the structure of the energy balance changes and environmental effects shift accordingly (Leitmann 1991). Also, the severity of impacts is affected by land use patterns: urban form and density influence energy consumption for travel; density and land use affect energy consumption for heating and cooling, and landscaping affects microclimatic conditions (Organization for Economic Cooperation and Development 1993).

2.11 ***Outdoor air pollution.*** Air pollution—closely linked to energy use—is a growing problem in large cities that have poor natural ventilation and high rates of motorization, industrialization, or coal use. In many cities, conditions are worsening as emissions from fuel use and industry increase (see Box 2.2). For example, vehicle fleets and associated emissions are expected to grow 5-10 percent a year in developing countries and most of the growth will be concentrated in major cities. Exposure to air pollution can result in breathing difficulties, hypertension, respiratory disease,

Box 2.2. Air pollution in selected urban areas

Bangkok: A recent study attributed rising levels of pollution to increasing population, industrial activity, and motor vehicle use. Two air pollutants that ranked in the high risk category are total suspended particulates and lead. The most important sources of particulate matter include industry, construction, transportation, fuel combustion, and trash burning. Particulate 24-hour emission (1,400 $\mu g/m^3$) is almost eighteen times more than the World Health Organization twenty-four hour guideline of 75 µg/m3 and could be responsible for up to 50 million restricted activity days and up to 1,400 deaths annually. Sources of lead include motor vehicles, industry, and smelters. Adverse effects from total exposure to lead could be responsible for many cases of neurological, cardiac, and other public health effects in both children and adults.

Cairo: The prevailing winds blow north or south; one brings toxic fumes into the city from the lead and zinc smelters in Shubra al Khaymay; when the winds shift, they bring poisonous pollutants from the steel and cement factories in the south of Helwan, an area known for its many dead trees. Helwan's cement plants emit 10,000 tons of dust per month into the air because of inadequate dust collection. The steel works and coke and chemical plant are also significant sources of dust emissions. Concentrations of airborne dust and dust fall exceed Egyptian standards by a factor of eight; it would require a 99 percent reduction in cement dust emissions to reach the U.S. standard of 0.3 kg/ton. Average levels of airborne dust have been as high as 400 $\mu g/m^3$ even though levels above 300 $\mu g/m^3$ are known to exacerbate chronic bronchitis and levels above 120 $\mu g/m^3$ increase the chances of respiratory illnesses among children.

Kuala Lumpur: The highly urbanized Kelang Valley, which includes Kuala Lumpur, has two to three times the pollution level of major cities in the United States. Kuala Lumpur has a concentration of suspended particulates in the air twenty-nine times the desirable goal recommended by the Malaysian Environmental Quality Standards Committee.

Katowice: Air quality in the Upper Silesia Industrial Region is very poor. Oxygen content is 18 percent on windless days in the most industrialized areas of this region and the carbon dioxide content is 800-850 percent above the standard. In addition, the average annual density of suspended particulates is fourteen times greater than the limit (314 $\mu g/m^3$ compared to 22 $\mu g/m^3$); sulfur dioxide exceeds the standard by three times on average; concentrations of benzo(a)pyrene exceeds the standard by ten to twenty-four times; and fluorine, formaldehyde, and phenol exceed the standards by three to eleven times.

Sources: Bangkok: Hauchmann (1991); Cairo: Khalifa and Mohieddin (1988); Kuala Lumpur: Sahabat Alam Malaysia (1983); Katowice: Borkiewicz et al. (1991)

impaired urological development, heart attacks, strokes, and death. Based on exposure to suspended particulate matter alone, rough estimates indicate that if unhealthy levels of particulates were reduced to the average yearly level that World Health Organization considers safe, between 300,000-700,000 premature deaths a year could be avoided in developing countries. This is equivalent to 2-5 percent of all deaths in those urban areas where levels of particulates are excessive (World Bank 1992). Urban transport-related air pollution is due to vehicle emissions, dust from unpaved roads, poorly-planned land use patterns, and other factors such as fuel quality, the age and condition of the vehicle fleet, poor traffic management, and atmospheric conditions.

2.12 ***Indoor air pollution.*** Distinct from ambient air pollution, indoor air pollution is possibly a more serious and ubiquitous problem throughout the developing world. A major household threat is the indoor burning of highly polluting traditional fuels, particularly for cooking and heating, which often results in daily exposure to high levels of toxic compounds. Smoke from the buring of biomass indoors contributes to acute respiratory infections that cause an estimated 4 million deaths annually among infants and children (World Bank 1992). Because exposure levels from indoor pollution often are quite high, those exposed can suffer respiratory infections that can lead to permanent lung damage and, eventually, heart failure. The problem is aggravated by high-density, poorly ventilated housing, the presence of primary and secondary cigarette smoke, and the use of toxic pesticides in homes. Occupational exposure to indoor air pollution, particularly to toxic emissions from small manufacturing enterprises, is another frequent problem. From a public health perspective, however, the key consideration is the total cumulative exposure of each person to pollutants from all urban sources—ambient, household, and occupational (Smith 1993).

2.13 ***Solid and hazardous waste management.*** In most developing countries, municipal authorities lack the resources and trained staff to provide their rapidly growing populations with the necessary facilities and services for solid waste management to support an adequate quality of life. Consequently, pollution from poorly disposed solid waste presents a formidable threat to health and productivity. In other situations, poor solid waste management may affect many people over longer periods of time. In Manila, for example, some 20,000 people occupy land surrounding a garbage dump known as Smokey Mountain. Children from informal settlements collect recyclable refuse from the dump or play amidst the garbage. The effects of such exposure on their health is not known. In April 1993 a solid waste dump site collapsed in Istanbul, causing the death of thirty-nine people whose homes were buried in garbage (twenty-seven were confirmed dead and twelve were missing) (Wilson 1993). Just seven weeks later, *The Washington Post* (1993) reported a similar "garbage slide" that occurred in San Salvador, killing twenty-two.

2.14 Poorly managed hazardous wastes present another growing threat to cities, particularly when industrial discharges are poorly regulated and when municipal waste management is inadequate. Because of these shortcomings, it is difficult to monitor discharges as well as to ensure that hazardous wastes do not end up in city sewers, landfills, or water used for drinking. This problem is compounded by the large quantities of wastes generated by small-scale and cottage industries and by hospitals and clinics located in and around cities. Human exposure to these wastes—whether inhaled, ingested, or absorbed through the skin—may result in short-term acute effects, long-term irreversible chronic diseases, or genetic mutations affecting future generations.

Resource losses

2.15 ***Ground water contamination and depletion.*** Improper disposal of urban and industrial wastes and urban runoff contributes to water quality deterioration of both surface and ground water resources. In the case of ground water, contamination by these sources is particularly serious and can lead to the permanent loss of valuable high-quality drinking water. Mexico City and Bangkok are extreme examples of what can happen when ground water depletion leads to land subsidence. Parts of Mexico City are sinking, breaking foundations and making some buildings uninhabitable. In Zelaya, Mexico, land subsidence has caused a 1.5 meter rift runs through the city. Subsidence is so serious in Bangkok that parts of the city are below sea level and floods plague the city every year during rainy season.

2.16 ***Land and ecosystem degradation.*** Environmentally inappropriate land development exerts direct pressure on land as well as on surrounding ecosystems. Particularly susceptible to inappropriate development are hillsides, floodplains, wetlands, coastal areas, and forests. For example, virtually all of the mangroves in Singapore have been reclaimed for urban development, resulting in the near-disappearance of coastal prawn ponds and fish traps that once provided abundant harvests (Chia 1979). In Sri Lanka, the draining of wetlands during the past fifteen years for urban development is posing serious flooding problems in parts of the Colombo urban area. Urban development also can have negative effects on downstream watersheds through increased runoff and erosion. In addition, watershed degradation can occur far from the city when major water supply or hydropower projects are built hundreds of kilometers away, or when firewood and charcoal is brought from long distances. In many cities, the occupation of hazard-prone land (for example, low-lying lands, floodplains, steep hillsides) by squatter and other illegal settlements is a significant cause of land degradation. In the case of Rio de Janeiro, extensive squatter development on high-risk hillsides, combined with deforestation and the dumping of solid waste in drainage canals, increases susceptibility to landslides that often accompany floods (Bernstein 1993). In Caracas, Venezuela, large portions of the city's poor live in barrios, or informal communities, on steep hillsides surrounding the downtown area. Landslides are common and can be devastating. They are caused by several confounding factors—slope instability, heavy rainfall, poorly built huts that cannot withstand subsidence and overcrowding (Diaz 1992).

2.17 ***Loss of cultural and historical property.*** The degradation of cultural resources (that is, archaeological and paleontological sites, historic and religious monuments, and natural wonders) and the loss of open space are often overlooked when considering urban environmental problems. Within existing built-up areas of cities, important historic structures or areas may be destroyed as part of government or private-sector efforts to establish high rise residential buildings, factories, or commercial centers in dilapidated downtown areas. In other instances, historic properties may be left to deteriorate while their municipal or commercial function is moved to newly developed areas of the city. Air and water pollution as well as the lack of infrastructure and building maintenance often accelerate decay and destruction. In Trujillo, Peru, squatters have invaded the adobe ruins of Chan Chan, the imperial capital of the ancient Chimú culture, and local authorities lack the needed resources to protect this priceless antiquity. Such damange affects ethnic pride and national identity and hurts tourism.

Environmental hazards

2.18 ***Natural hazards.*** Many urban areas are subjected to severe or frequent floods, wildfires, earthquakes, tropical storms, volcanic eruptions, or other catastrophic natural events. When such events occur in or near densely populated areas, the damage can be disastrous. The costs of the 1988 flood in the Rio de Janeiro metropolitan region exceeded $900 million. In metropolitan Buenos Aires, damage due to heavy flooding in 1985 was estimated at over $200 million. In most countries, inadequate planning, insufficient infrastructure and services, inappropriate building design, faulty construction, inadequate maintenance, and overcrowding all contribute to both environmental degradation and increased vulnerability to such catastrophic events (Kreimer and Munasinghe 1991). For example, the inhabitants of coastal areas are more vulnerable to floods, typhoons, or tsunamis when land reclamation projects encourage settlement in dangerously low-lying areas or when land clearing, construction, and harbor dredging remove protective vegetation, reefs, or sand dunes.

2.19 ***Man-made hazards.*** Such man-made phenomena as severe pollution episodes, chemical contamination, and industrial accidents constitute both acute and chronic environmental hazards. Their destruction tends to be the most profound in cities where squatter settlements or illegal subdivisions are concentrated around industrial and commercial areas or hazardous waste disposal sites. Severe events such as the 1984 disaster in Bhopal involve thousands of lives and the razing of homes and industrial facilities. Since the urban poor tend to settle near industrial facilities (as in Bhopal) or build their homes on vulnerable land, they are extremely vulnerable.

2.20 In 1991, five people died and over 3,000 were left homeless by fires from a massive chemical explosion in Bangkok's Klong Toey port (Shin et al. 1991). Large sacks of phosphorus caught fire, then several 200-liter tanks of liquid alcohol and three tanks of chemical gas exploded. As a result, three warehouses owned by the Port Authority of Thailand storing toxic chemicals and other chemicals were destroyed. Reported health effects included blood levels of methyl bromide above the acceptable safe levels, congested breathing, sore throats, and burning pain on the skin resulting from subsequent rainfall. Damage to goods and warehouses totalled more than $1.55 million. In 1992, a tanker truck in Dakar, Senegal, filled with liquid ammonia exploded at a peanut-processing factory, killing at least sixty people and injuring about 250. The explosion, which was the worst industrial accident in Senegal, occurred during working hours. At the time of the explosion, a shortage of gas masks made rescue efforts difficult because of toxic ammonia gas. Also, in 1992, a series of powerful explosions caused by liquid hexane dumped in the municipal sewer system killed at least 212 people and injured 1,000 others in Guadalajara, Mexico. In addition to the deaths and injuries it caused, the explosion destroyed homes, crushed small buildings, and tore up streets (*The New York Times* 1992).

2.21 With chronic exposure to hazards, as in the case of Katowice, Poland, rates of cancer are often high and lower life expectancies are linked to environmental conditions. The city, which in 1990 had a population of 373,000, is the industrial center of the Upper Silesian Industrial Region and the capital of the Voivodeship governate. The Katowice municipality has a heavy industrial base—large industrial plants, coal mines, metallurgical plants, and over 2,000 factories producing processed food, chemicals, and other equipment that contribute to severe pollution. According to Hertzman (1990), urban life expectancy in the Katowice Voivodeship is the shortest in Poland; for men and women, it is more than a full year lower than the national average. Urban male life expectancy also has fallen over the past fifteen to twenty years. Children have been especially vulnerable to the environmental conditions in the area. In Katowice City, the 1989 infant mortality rate was 25.5 per 1,000 live births, which compares unfavorably with the national average of 16.1 per 1,000. The results of a six-year study during the 1980s in sub-regions of the Voivodeship conclude that infant mortality is correlated with dust fall, ambient level of lead, tar, phenols, formaldehyde, and benzo(a)pyrene. The Voivodeship also has the highest incidence of premature births (8.5 percent), genetic defects (10.1 percent of all live births), and spontaneous miscarriages. Chronic bronchitis is reported in 35 percent of children living in heavily polluted industrial areas.

Factors Aggravating Urban Environmental Degradation

2.22 To reverse urban environmental degradation in most developing countries, it is essential to understand and specify the factors that perpetuate the lack of appropriate preventive and curative environmental actions. In large part, urban environmental problems can be attributed to institutional deficiencies, inadequate policies, and actions (or inaction) by public and private actors. Among the prevailing factors are the absence of full participation, inadequate governance, inadequate regulatory and economic policies, and insufficient knowledge and information.

Lack of public and political awareness

2.23 One of the main contributions to environmental degradation is the lack of public awareness of the problems and low participation in efforts to improve the urban environment. Sometimes the link between environmental cause and effect is so remote that it is not recognized, or if it is, social and private interests may diverge, with little inclination to participate. (Participation here refers to an active, sustained role in influencing policy formulation and other key decisions.) Political leaders often focus on immediate and highly visible problems, leading to short-term "band-aid" solutions and are inclined to skimp on meeting the recurrent costs of maintaining local infrastructure or the investments needed to control environmental spillover effects that extend beyond political boundaries. Regrettably, it often takes an environmental disaster to stimulate profound change. The chemical accident in Bhopal, India (mentioned earlier) is a powerful example of an environmental disaster in which political expediency was a contributing factor leading to failure in enforcing industrial safety and land use control regulations.

2.24 ***The need for public pressure and political will.*** In the absence of public pressure to improve environmental controls over industry and other polluting sectors, the easiest government response is inaction or disregard for environmental considerations. In making investment decisions, government policymakers typically consider direct economic benefits above all other considerations. In assessing a proposed industrial facility, for example, the prospect of new jobs usually overrides concern about potential air or water pollution. Another impediment to industrial pollution control is that many of the large-scale pollution-intensive industries (for example, mining, paper and pulp, chemicals, iron and steel, and nonferrous metals) are owned and managed by the state or by the politically powerful upper-income elite. Under these conditions, government officials have a difficult time summoning the political will needed to impose strict regulations, particularly when the targeted industries are viewed as vital to economic development.

2.25 ***Lack of effective public education and participation.*** To apply pressure on government officials as well as polluters to improve environmental conditions, citizens need to be educated about local environmental quality, the effects of existing environmental management practices, and available alternatives. In many countries, however, environmental education is not included in the formal educational system. In other instances, well-conceived environmental education or public information programs are sometimes undermined by the media, who tend to sensationalize environmental topics rather than focus on real local priorities. Further, most local authorities fail to involve the public in the design of environmental improvements. This undermines the chances that such reforms will succeed and be maintained in the long term. When users are not consulted in the design of local services, they normally are less willing to pay for new or improved services than when they are involved in system planning and design.

Inadequate governance

2.26 Inadequate governance (where "governance" refers to the exercise and sharing of power) is a principal constraint to effective urban environmental management. In most developing country cities, metropolitan and municipal governments lack the institutional capacity to carry out effective environmental planning and management and to routinely provide effective urban services. Other key impediments to effective governance include poor coordination when public responsibilities of different levels of government overlap, and the failure of most governments to provide adequate public education or include community and private sector participation in the design, planning, and/or implementation of environmental services. An overriding constraint is the lack of effective public accountability—citizens are often unable to voice their concern about the quality and accessibility of services and, where service is inadequate, they may not be able to opt for alternative sources of supply.

2.27 ***Weak institutional capacity.*** Weak institutional capacity hampers most efforts to improve environmental conditions in cities. It is characterized, for example, by unclear planning and operational responsibilities and by poorly executed municipal waste management and pollution control programs. Similar shortcomings affect such areas as urban land management, traffic control, and municipal finance. In many countries, separate sector agencies plan for and provide individual environmental services, but coordination between agencies is inadequate. Although inefficient management of municipal and industrial waste treatment and disposal operations makes pollution control a major priority for most urban regions, public agencies lack the necessary autonomy and resources to make the provision of urban environmental services more efficient. One reason for this is that public agencies may expend little effort to involve the private sector. In many developing countries, there is no urban environmental planning, or where it exists, it fails to achieve broad-based consensus on environmental goals and objectives. Also absent are efforts by the local authorities to incorporate appropriate technologies, community involvement, and where feasible, private sector (both formal or informal) participation. Efforts to mobilize financial resources, such as through property taxes and user charges, are often inadequate. This inability to raise funds is an important factor accounting for the failure of local authorities to cover the full costs of operation and maintenance.

2.28 ***Jurisdictional complexity.*** Poor coordination among the institutions responsible for environmental management and for urban management and services presents another constraint to effective environmental policies. Environmental management responsibilities can involve national and regional sector agencies, state or provincial governments and competing local governments. These overlapping responsibilities, together with the competing priorities of local neighborhoods, district government, municipal officials, and regional, state and national policymakers, contribute to jurisdictional complexity. Resolving issues of jurisdictional complexity and spatial scale (that is where the environmental problems spill over local boundaries), requires finding a balance between decentralized and centralized approaches. This involves finding the comparative advantages of both national and local institutions in carrying out different aspects of urban environmental management activities. National agencies, for example, are better equipped to set emissions standards or resource pricing policies while local agencies are better-suited to provide municipal services.

When policies don't measure up

2.29 In most developing country cities, regulatory and economic policies have not provided the necessary foundation for effective urban environmental management. Many countries have not established realistic laws, regulations, standards, and enforcement mechanisms to cope with the most basic environmental problems. Similarly, existing economic policies in many developing countries have actually encouraged waste and pollution through inadequate pricing and other sources of economic distortion.

2.30 ***Inadequate regulatory policies.*** Only a few countries have established comprehensive and realistic environmental protection laws and regulations. Existing pollution control laws in many countries do not clearly assign jurisdiction and responsibility for monitoring and enforcement, nor do they specify funding mechanisms for control and enforcement. Such laws and regulations sometimes are not consistent with current scientific knowledge and fail to establish clear objectives and policies. Steps for issuing new regulations and standards are not spelled out, and guidelines for how to set sanctions equitably are not provided. Moreover, excessive regulations and standards that are costly and politically unpalatable sometimes hinder realistic progress in pollution control. In many instances, however, significant environmental improvement could be achieved with existing laws and regulations, provided there is sufficient monitoring and enforcement. In countries where the environment is not a political priority, this problem is acute.

2.31 Inappropriate urban land use regulations are probably the most important factors accounting for losses of ecologically and culturally important resources and for the location of low-income populations on hazard-prone lands. Land use regulations are often unaffordable, difficult to implement, or unenforceable. In some instances, the problem lies in excessive regulation whereby land development standards artificially reduce the supply and raise the price of land by requiring large lot sizes or excessive amounts of land for circulation or green spaces. In other situations, governments have not formulated effective land use policies, laws, and standards that deal with development in sensitive areas and hazard-prone lands or that guide urban expansion away from areas poorly suited to urban development.

2.32 The extent of the tragic 1984 industrial accident at Bhopal in India was due in part to poor land use controls. Because of a severe housing shortage and a lack of zoning enforcement, 156 illegal settlement slum communities grew up around the pesticide plant, two of which were across the street from the Union Carbide facility. Close to 3,000 people died in the accident and more than 300,000 suffered exposure-related ailments.

2.33 ***Unclear property rights.*** Unclear property rights contribute to the degradation of land and natural resources as well as to the inequitable distribution of these resources. Few countries have resolved basic land tenure problems that impede mobilization of household and community resources for basic urban environmental improvements. This is especially the case for peri-urban squatters and central-city slum dwellers. The pervasive lack of secure tenure and inadequate land titling registration systems have a profound effect on the ability of the poor to acquire safe land for housing. Without clear property rights, landowners are unable to gain access to formal credit because they cannot use their property as collateral. Moreover, the absence of secure tenure leaves residents with little incentive to maintain their dwellings or to invest in improvements that will reduce their vulnerability to floods, earthquakes, and other natural hazards. Where water rights are unclear, the extraction of ground water for urban and industrial purposes often is in direct conflict with the efficient development of regional water resources.

2.34 ***Inefficient economic policies.*** Inefficient economic policies include distorted pricing of energy, natural resources, and urban services; ineffective taxes; and wasteful subsidies. Inefficient pricing of such essential resources as land, water, energy, and food undercuts attempts at resource management and promotes excessive use of scarce natural resources, which leads to eventual ecosystem collapse. In Algeria, Egypt, Turkey, and Yugoslavia, for example, water prices are usually 20 percent of marginal costs (World Bank 1993). Consequently, industries and households use greater quantities of water than are needed for essential uses. This generates excess wastewater, which in most cases is discharged untreated into surface water. Excess water use also may trigger the construction of unnecessarily large treatment plants and other facilities. Inappropriate pricing of urban services results in inefficient allocation and inequitable distribution—the urban poor are left without public services (sometimes despite their willingness and ability to pay); services for high income groups are subsidized; capital and recurrent expenditures are misallocated; and local costs are not recovered. These inefficiencies at the local or municipal level lead to greater dependence on fiscal transfers from the central government.

2.35 Hidden subsidies and poorly devised or enforced taxes, interest rates, import and export policies, exchange rates, and regulations contribute to the distorted use of environmental resources. Protective tariffs can encourage the manufacture of energy-inefficient or highly-polluting products. The absence of transport charges or stumpage fees for urban-bound fuelwood or for charcoal production, for example, leads to ineffective forest management. An artificially low interest rate for municipal loans or housing finance (set far below the cost of funds) undermines the sustainability of the financing mechanism. Excessive land use controls result in unaffordable urban land prices, which leads to environmentally inappropriate development (for example, on cheaper land that may be vulnerable to natural disasters or unacceptably close to a high-polluting industrial area).

Insufficient knowledge and information

2.36 The capacity to plan and implement responses to urban environmental problems, and the ability to build a political consensus, is hampered by insufficient knowledge of both the problems and the processes of environmental degradation. A recent global review of urban research in the developing world (Ford Foundation 1993) found that, although virtually every regional analysis of urban research emphasized the urban environment as a priority topic for the urban research agenda in the 1990s, there was scant evidence of actual research having been completed and disseminated. In a review of all developing regions, only about 2-3 percent of the urban research had an environmental focus, notwithstanding the stated priority of the urban environment.

2.37 While adequate environmental data for urban areas are usually lacking, so are accepted analytical frameworks for understanding the problems, ranking them, and designing locally appropriate environmental protection programs to resolve them. For example, most authorities are not aware of the magnitude of ongoing environmental damage or how various waste disposal practices may be threatening human health and environmental resources. Similarly, most local authorities do not have adequate data on the locations of environmentally fragile or hazard-prone lands. They have not pinpointed where vulnerable populations are, nor do they maintain adequate records on the locations and capacities of existing utilities and services. Most cities also lack the necessary information to facilitate land transactions, effective land use planning, property taxation, and hazard mitigation. To make matters worse, education programs covering the scientific, technical, and managerial aspects of urban waste management and pollution are either weak or non-existent. Consequently, there is a shortage of professionals who can adequately understand and analyze the relationships between environmental problems, impacts, causes, and preventive and curative actions.

Defining Environmental Impacts and Setting Priorities

2.38 Given the broad range of political problems and their causes, it is necessary to set priorities. Priority setting involves valuing the impacts of urban environmental problems and initiating an informed participatory process to agree on the best issues and options.

Valuing urban environmental problems

2.39 The challenge of environmental planning is to value the effects of and rank urban environmental problems in terms of health effects, productivity, amenity, ecological values, and other key indicators. By categorizing impacts in this way, emphasis can be placed in the following areas:

- *Health and Safety.* In examining health and safety effects, assessment criteria include health care costs, lost working days, and higher mortality rates.

- *Productivity.* Any assessment should judge the extent of losses in urban productivity, which is the result of poorly run cities and such external factors as spillover pollution from nearby industrial areas.

- *Equity.* Because the negative effects of urban environmental degradation fall disproportionately on the poor, issues of equity are of critical importance. Efforts to improve universal access to basic sanitation and other essential urban services and to include the poorest of the poor in urban improvement programs will go a long way toward mitigating urban environmental degradation.

- *Ecology.* Ecological effects can be judged by the availability and costs of fresh water, the extent of land subsidence from overdrawn ground water, the vulnerability to flooding, landslides and other natural disasters, and the loss of biological diversity.

- *Amenity.* Effects on amenity include air and water quality, noise levels, scenic beauty, and the presence of parks and clean, open spaces.

Setting priorities—lessons of experience

2.40 Efforts to establish environmental priorities (by conducting rapid assessments and issue-specific action plans as described in chapter 4) are under way in many cities. These efforts reveal several environmental priorities and common concerns. Assessments are completed or are under way in Bombay, Colombo, Dar es Salaam, Manila, Mexico City, São Paulo, Abidjan and seven secondary cities in Côte d'Ivoire, and five cities along the Mediterranean. Case studies of several of these cities are presented in the Annex.

2.41 An effective way to rank problems and priorities is to build a typology divided into externalities and impacts. Such an approach is especially helpful when confronted with a wealth of information—either across several sectors or from several cities (such as when assessing a metropolitan area or when conducting a regional urban study). The main externalities and impacts affecting cities include those arising from inadequate environmental hygiene and waste management (pollution), from the congestion of urban systems, and from degradation of natural support systems. The major impacts associated with these environmental problems relate to health and safety, productivity, ecology, amenity, and equity.

2.42 Several valuable lessons have emerged from a study on valuing the economic impacts of environmental problems in Asian cities (Shin et al. 1993). After surveying what is known about urban environmental problems and their costs in the region, the study assesses approaches to the economic valuation of environmental effects, discusses alternative valuation measures, and makes suggestions for applying valuation techniques in the near future to selected Asian cities.

2.43 Based on the typology described above, the Asian study concludes that the major impacts on health and safety come from air and water pollution—two issues that merit urgent attention. According to the study, productivity is most affected by traffic congestion and water pollution. Amenity value is adversely affected mostly by air and water pollution and traffic congestion. Losses of critical ecosystem functions and of biodiversity have become irreversible in many cities. Finally, the study confirms that these impacts, particularly on health and safety, fall most heavily on the urban poor because they lack access to basic environmental services. Thus, the search for solutions in cities in developing countries should be based on the disaggregated analysis of impacts on the poor and non-poor populations (Shin et al. 1993).

2.44 Another approach is to conduct a rapid urban environmental assessment to compensate for the gap between available research and the need for action at the local level. The Urban Management Programme used the rapid assessment approach in an effort to close the gap in information on the urban environment in developing country cities (Leitmann 1994). To obtain the information and carry out the assessment, researchers used an urban environmental questionnaire, an urban environmental profile, and a consultation process (described in chapter 4). Results were acquired from six cities and one urbanizing area. Included were Accra, Jakarta, Katowice, São Paulo, Tianjin, Tunis, and the Singrauli region. The following broad conclusions regarding priorities can be drawn from the preliminary assessments:

- Urban environmental strategies should focus explicitly on the problems of the poor.

- City-specific strategies should be guided by a range of economic considerations, from the potential for cost recovery, to income level, to the viability of enforcing pollution taxes or user fees for improved environmental services.

- Solutions that are not heavily dependent on institutional performance may be needed in the short-run because of entrenched organizational problems that require medium- or long-term solutions (particularly those compounded by cross-jurisdictional issues).

- Enhanced public awareness, consultation, and participation can improve environmental management.

- When designing interventions, careful attention must be paid to the selection of problem areas, their scale, and available institutional capacity.

Priority Environmental Problems and Policy Linkages

2.45 Once a thorough assessment is made of the most pressing environmental problems in a city and their underlying causes are pinpointed, such information should be used to devise policy responses, or, at minimum, to study the policy implications of the problems. For this purpose, a matrix linking key policies to problems would be useful for economic, environmental, and urban planners and decisionmakers, as in Table 2.1. In chapter 3, the policy areas highlighted in the table will be examined and specific policy options will be discussed.

Table 2.1. Critical policy links for urban environmental management

Urban environmental management issue	Underlying causes	Relevant policy reforms
Access to basic environmental infrastructure and services:		
• Serviced land and shelter	Poorly functioning urban land and housing markets; Highly regulated prices; Lack of affordable housing for the poor	Reform property rights; Develop mortgage financing; Introduce affordable standards and target subsidies to the poor; Reduce unneeded regulations, government interventions and subsidies
• Water supply, sanitation, drainage, solid waste collection, transport	Supply side dominated by government monopoly; Prices heavily regulated; Heavy subsidies	Introduce pricing and demand management; Reduce subsidies; Move toward decentralization, privatization, participation
Pollution from urban wastes and emissions:		
• Water pollution	Uncontrolled municipal and industrial discharges; Excessive water use and waste generation; Failure to link water quantity and quality issues	Introduce water pricing and effluent charges; Subsidize sewage treatment; Strengthen regulations and capacity for monitoring and enforcement; Prepare comprehensive basin plans
• Energy use and air pollution – ambient air pollution – indoor air pollution	Increased motorization and transport congestion; Energy supply side dominated by government monopoly; Heavy energy subsidies; Household and cottage industry use of low-quality fuels	Introduce energy and fuel pricing, road charges, emissions charges; Reduce automobile subsidies, fuel subsidies; Integrate transport and land use planning; Promote clean technologies, fuel substitution, vehicle maintenance
• Solid and hazardous waste management	Poor municipal management; Lack of disposal facilities; Inadequate regulation and enforcement	Introduce regulations, licensing, and charges; Stimulate waste minimization; Strengthen municipal management, operations; Privatize disposal operations
Resource losses:		
• Ground water depletion	Unsustainable extraction linked to unclear property rights and treatment as free resource	Clarify property rights; Introduce extraction charges
• Land and ecosystem degradation	Low-income settlements "pushed" onto fragile lands by lack of access to affordable serviced lands (see above); Lack of controls over damaging economic activities	Coordinate land development; Remove artificial shortages of land; Develop sustainable uses of sensitive areas; Monitor and enforce land use controls
• Loss of cultural and historic property	Lack of property rights, regulations, enforcement, maintenance; Failure to reflect social values in land prices	Introduce tax incentives for preservation; Use redevelopment planning, zoning, and building codes; Develop property rights

continued on next page

Table 2.1 continued. Critical policy links for urban environmental management

Urban environmental management issue	Underlying causes	Relevant policy reforms
Environmental hazards:		
• Natural hazards	Poorly functioning land markets; Ineffective land policies; Poor construction practices	Enable land markets (see above); Provide disincentives to occupation of high-risk areas, incentives for using disaster-resistant construction techniques;
• Man-made hazards	Inadequate regulation and enforcement; Low-income settlements alongside hazardous activities	Introduce and enforce environmental zoning; Formulate urban disaster preparedness plans and strengthen response capacity

III. OPPORTUNITIES TO IMPROVE THE URBAN ENVIRONMENT

3.1 Options for confronting urban environmental degradation correspond with its basic causes and include efforts to focus on cost-effective approaches, to mobilize public support and participation, to improve governance (which encompasses institutional capacity-building and better provision of urban services), to improve policy interventions, and to close the knowledge gap. Mobilizing public support and participation for environmental management involves raising awareness of problems and solutions, building constituencies that will demand an improved quality of life, and promoting participation by NGOs and by the informal sector. Upgrading the management and delivery of key urban services is also essential. On the policy side, efforts should focus on improving the coverage, efficiency, and effectiveness of individual interventions as well as on choosing an appropriate balance of policies and instruments. This can be done efficiently by creating mutually supportive packages that emphasize strategic investments and financing.

Improving the Urban Environment: Key Principles

3.2 Once environmental priorities are defined, the selection of urban environmental policies should be based on the severity of the problem, on its economic and health costs, and on the available instruments to carry out the reforms. Because of budget constraints and other competing demands for scarce resources, most cities in developing countries will achieve the greatest success if they seek out cost-effective solutions that focus on one or two priority problems. When designed effectively, environmental policies strongly influence the behavior of both public and private actors, thus affecting urban environmental quality. For example, such instruments as user fees or taxes for urban services affect waste management, water use, and other household and industry practices. Similarly, regulations and enforcement systems affecting urban land markets influence the extent to which environmentally fragile land is converted to urban use.

Seeking "win-win" situations

3.3 Win-win environmental solutions occur when environmental and economic goals are complementary. One such examples is Mexico City's environmental action plan for improving air quality. In the megacity, where the economic damages due to the health effects of air pollution are estimated to be $1.5 billion a year, the World Bank and the Mexican Government have developed a win-win plan for air quality management. By comparing the cost per kilogram of emissions eliminated through options ranging from vapor recovery to fuel improvements, Bank researchers were able to rank the cost-effectiveness of several alternative measures. They found that significant reductions in transport-related emissions are possible at moderate cost by using a combination of regulations, incentives, and fuel taxes. Based on the least-cost analysis, the Bank is supporting a transport and air quality management project for the city. It includes a vehicle component that will support the replacement of old high-use vehicles (such as minibuses) with new emission-controlled vehicles as well as funding for fuel conversions and a retrofit of emissions controls. Also included are components related to fuel, transport management, air quality research, and monitoring and institutional strengthening (see Eskeland 1993 and Case Study 6 in the Annex).

Cost-effective approaches

3.4 The level of spending required to carry out environmental reforms and the benefits that will accrue to targeted communities are of critical importance. For example, municipal solid waste management usually accounts for a third to a half of available municipal operating budgets. Capital investments for environmental improvements such as municipal wastewater treatment can run as high as several hundred million dollars. For most cities, however, the costs of resolving the principal environmental problems will be affordable, with overall benefits exceeding costs. The cost of gradually phasing in lead-free gasoline, for example, is relatively inexpensive. Even where benefits are not easily quantified, if there is broad public and political agreement on environmental priorities then the decision rule should be to undertake the most cost-effective solution that meets the environmental goals.

3.5 Given the damage caused by inadequate environmental management, cities have no choice but to invest in environmental improvements. In Jakarta, where water supply is inadequate, more than $50 million (an amount equal to 1 percent of the city's GDP) is spent each year by households to boil water—a tremendous energy expense. Investments in water supply can therefore reduce fuel wood consumption and air pollution (World Bank 1992). In Metropolitan Manila, the implementation of the proposed environmental management strategy project, which includes investments in wastewater treatment, reforms to industrial waste management policies, vehicular and industrial emissions monitoring, institutional strengthening, and many other measures will cost $186 million. Without the project, the estimated annual economic and social costs of continued urban environmental degradation in Manila would be about $53 million in the year 2000. By that time, the annual net environmental benefits associated with the project will be an estimated $32 million (see Case Study 3 in the Annex).

Stressing economic efficiency and cost recovery

3.6 For environmental management strategies to work, increased capital investment (to meet the growing backlog of urban environmental infrastructure needs) must go hand-in-hand with more cost-effective technologies for environmental protection, and with greater economic efficiency and cost recovery. Determining and responding to the effective demand for urban services should be the basis for decisionmaking. Reliance on the public budget for such investments should be reduced by applying user charges, property and business taxes, and fuel taxes. The viability of an intervention will depend on matching its cost to users' ability and willingness to pay; potential users should be ready to pay for the full cost of improved services, including the maintenance and replacement of necessary equipment. Even for such basic services as waste disposal, self-financing should at least be sufficient to pay for recurrent costs.

3.7 In practice, user charges have been applied in both developed and developing countries with varying degrees of success. In some parts of France, for example, individual households pay the water basin agency a pollution charge and the local community is responsible for collecting and treating wastewater. This willingness to pay for improved basic services also extends to the developing world. For example, in West Point, an informal settlement in Liberia, payment for the use of sanitary facilities covers operating and maintenance costs. In Bihar and Delhi, India, user fees have allowed Sulabh International (a non-profit organization) to extend the benefits of pay-and-use latrines to many urban neighborhoods. Ideally, user charges should reflect the incremental costs of building, maintaining, and operating environmental infrastructure. Although user charges can reduce government costs for waste management, public investment may still be needed, particularly for the environmentally sound disposal of urban wastes.

Mobilizing Public Support and Participation

3.8 Improving the effectiveness of environmental decisionmaking requires sustained participation by a diverse cast of actors. Through participation, people (especially the disadvantaged) can influence policy formulation, design alternatives, investment choices, and management decisions (Bhatnagar and Williams 1992). They can also play a role in the monitoring of environmental interventions in their communities. Thus, participation should occur continually over the lifecycle of a program or project. Three groups of actors should participate in planning and management:

- those whose interests are affected by environmental problems, strategies, and plans;
- those who control relevant implementation instruments; and
- those who possess relevant information and expertise.

Of these three groups, the first is the most important for accomplishing change. Urban environmental improvement cannot be initiated or sustained without constituencies that demand a better quality of life. Participation often begins when concerned actors apply political pressure for environmental action. To lobby effectively, the most vulnerable urban groups (particularly the impoverished, women, children, and the elderly) must understand the causes of their problems and articulate their concerns about the planning and implementation of environmental infrastructure and services. The second and third groups of actors are needed to mobilize this participation and to improve environmental management. Mobilizing broad support requires efforts to raise awareness, especially among vulnerable groups. It also requires constituency-building and the promotion of NGO and informal sector participation.

Raising awareness

3.9 To gain awareness, the public and the cast of actors involved in urban environmental management need to learn about environmental options, solutions, enforcement, and monitoring. This can be accomplished through public and private agencies and by disseminating information about environmental risks and how to mitigate or avoid them. The goals of awareness-building are to motivate affected groups to participate in environmental management, stimulate grass roots programs to improve and protect the urban environment, and promote innovations and local know-how for effective organization and management. The mechanisms for raising awareness can be formal (for example, school courses on ecology and urban life), or informal (such as monitoring of pollution by neighborhood groups). Options that have been used in different cities include:

- *Formal community education* to stress the linkages between urban sanitation and public health (this is taught through the Free University of the Environment in Curitiba and is called *pendidikan masyarakat* where it is taught in Jakarta and other Indonesian cities).

- *Ecology courses* as part of the formal school curriculum (also used in Curitiba and Jakarta).

- *Resource conservation campaigns* to promote the efficient use and re-use of fuels, water, paper, metals, glass, plastic, and other resources (such as the Quezon City Cleanliness and Environmental Awareness Drive in the Philippines).

- *Targeted environmental education* for community leaders, politicians, city officials, and others (as in Tianjin and Jakarta).

- *Regular, publicized evaluations and awards* for environmental improvement and urban environmental quality (such as through the Environmental Responsibility System in Chinese cities, through ADIPURA (Clean City) awards for Indonesian cities, and through specific awards for communities in Jakarta).

- *Media campaigns* that research and publicize particular issues of community concern (such as the radio station-sponsored to clean up the Tiete River in São Paulo and the Magic Eyes campaign to clean up Bangkok's Chao Phraya River).

- *Regular information on environmental quality* such as the *Alerta 2* monitors in São Paulo, and air pollution reports in the Mexico City news media.

Building constituencies

3.10 In most Third World cities, a powerful constituency for the environment already exists among the upper class. They usually are most concerned about issues relating to green/recreational space, preservation of low-density residential areas, and continued access to the basic services and infrastructure that benefit them. The challenge, however, is to build a constituency of urban poor. These interest groups can be built from the bottom up or from the top down. In the bottom-up approach, groups of concerned citizens band together to demand improvements in environmental quality, either from public officials, private corporations, or both. For example, in many squatter settlements, organizations are formed to apply pressure for regularizing neighborhoods by lobbying for the provision of water supply, sanitation, and drainage. In the top-down model a politician will build support by promising to upgrade environmental services in exchange for votes at the next election. Alternatively, pressure may come from both levels, as in the case of Olinda, Brazil (Box 3.1) where initial political patronage resulted in the creation of a community organization that lobbied for environmental improvements.

Involving NGOs and the informal sector

3.11 Local participation for managing the urban environment can take the following forms: (i) information-gathering on community conditions, including needs and impact assessments; (ii) articulation of and advocacy for local preferences and priorities; (iii) consultations concerning programs, projects, and policies; (iv) involvement in the selection and design of interventions; (v) contribution of "sweat equity" or management of project implementation; (vi) information dissemination; and (vii) monitoring and evaluation of interventions.

Box 3.1. Creating a constituency for the environment in the Olinda slums

In 1983, the municipal government of Olinda (a city of 400,000 in the metropolitan area of Recife, Brazil) began a project for environmental upgrading, starting with a pilot effort in the Triangulo do Peixinhos slum. Olinda consists of a small historical center (a UNESCO World Heritage site), bounded on one side by a middle-class neighborhood, and on the other by a large slum. Triangulo, part of the larger Peixinhos slum, is a well-established low income community housing about 2,400 people. The pilot activities have been sustainable, in part, due to a high degree of community participation.

Initially, the project focused on improving **drainage**, using low cost fiber-crete drains (US$1 per meter constructed) and land fill. The employees who carried out these works were hired through the local community association. In response to community pressure, the project was expanded to include **solid waste collection.** Workers were again contracted in the community to collect household refuse using handcarts. A small aerobic composting unit (250 m^2) was established at the former site of an illegal dump to handle all domestic waste in the neighborhood; the neighborhood then gradually developed as a center for recycling businesses. Next, the project became involved with **sanitation** through the construction of ventilated improved pit latrines (each constructed unit cost $90; 25 percent cost recovery was achieved via repayment in an inflation-proof currency—five bags of cement per latrine), and **road upgrading** with low cost soil/cement paving (US$1.50/$m^2$). All components were produced locally through newly-created micro-enterprises.

Community participation was vital to the success of this venture. First, as mentioned, labor was contracted directly from the area through the community association. Neighborhood residents were responsible for maintenance—for example, regular sweeping of the drainage gutters with proper storage of the collected waste for removal by the municipal refuse service. While initial community concerns were voiced through a larger neighborhood association and several political leaders, project activities stimulated the creation of a Triangulo dwellers' council. The council then began to articulate the residents' priorities for environmental improvement. The city supported community involvement by creating a Special Projects Unit, as part of the municipal urbanization company (URB-Olinda), that provided training and public education in addition to urban upgrading.

The success of the Triangulo pilot activity led to its replication in other areas of Olinda: Ilha de Santana, the historical area (low cost sewerage, decentralized solid waste management), and the larger Peixinhos slum (drainage). While this replication has been discontinued more recently due to lack of political support, residents of the Triangulo area have continued their environmental management efforts. This sustainability is attributed to three factors: (a) the regularization of community involvement through direct contracting; (b) development of community awareness about the importance of maintaining environmental infrastructure; and (c) the mobilization of household resources for local investment that often accompanies government-funded community upgrading.

Source: Interview with Ruy Rego, former technical coordinator of the Triangulo do Peixinhos project, and former head of URB-Olinda, April 1992.

3.12 Where formal public participation in government operations has been lacking, three alternative approaches have been used to stimulate public involvement in urban environmental management. First, NGOs or CBOs often champion local concerns related to environmental management issues. Second, consultations and town meetings are often used as a forum to discuss environmental priorities, constraints, and opportunities as a prelude to concerted action. In Dar es Salaam, city consultations were held recently between local actors affected by urban environmental problems, national ministers, and donors. They met to clarify urban environmental goals and set program priorities (See Case Study 1 in the Annex). Likewise, the Government of Côte d'Ivoire is pulling in multiple actors from all levels through consultations to set priorities for an urban environmental action plan that will involve Abidjan and seven secondary cities. Town meetings will also be held (see Case Study 2 in the Annex).

3.13 Where CBO/NGO representation or consultations have not been possible, a third avenue of direct action has been pursued. This can take three forms—protest, legal action, or grassroots management. When an environmental problem becomes intolerable and no plausible recourse is possible, communities may turn to public protest, ranging from peaceful to violent. The case of Cubatao is an example of the potentially important role that public pressure has in reversing urban environmental degradation. In this Brazilian city, industrial air pollution was allowed to reach levels that caused genetic defects and impaired lung function in children, spontaneous abortion, reduced life expectancy, and other health effects. These problems were addressed only after outraged citizens organized and demanded action. Public outcry against polluting facilities and toxic discharges may even become violent, such as in 1986, when a new tantalum production plant was burned by rioters in Phuket, Thailand (Wheeler 1992).

3.14 Legal action is another avenue for public involvement. For example, in July 1992, residents of Bukit Merah in Malaysia obtained an injunction to shut down a Japanese-run chemical firm that was dumping thorium hydroxide, a radioactive byproduct capable of causing genetic defects and other ailments. In Mexico, legal action by citizens groups is possible through newly created environmental courts. Faced with an unresponsive or repressive system, some communities take matters in their own hands. Such initiatives have included: (i) local monitoring and reporting on polluting industries in São Paulo; (ii) the provision of environmental services in the San Juan and Valenzuela municipalities of Metro Manila; (iii) improvements to streets and sidewalks in the barrios of Buenos Aires; (iv) the provision of low-cost sanitation in northeast Brazil and in Karachi (Box 3.2); and (v) the establishment of health centers and health-related services in Jakarta's neighborhoods through the *posyandus* (integrated health posts). In a well-managed city, beneficiaries participate by paying user fees on a regular basis and by applying political pressure when services are deficient.

Improving Governance—Building Institutions and Improving Urban Services

3.15 The tell-tale signs of a government in need of improvement are weak institutional capacity and inefficiently- or poorly-run urban services. In this section, improving governance encompasses capacity-building at the local level, choosing the tools for capacity-building, clearly delineating institutional arrangements, and improving the operation of urban services.

3.16 Strong institutional capacity is a critical pre-requisite for improving urban environmental problems. Although developing country cities are carrying out various urban environmental management activities, many will need to upgrade the institutional and technical capacities of key actors to help identify, understand, and evaluate complex urban environmental problems. Key actors must also have the skills needed to establish priorities, resolve conflicts, and design and implement appropriate policies and investments. In laying out a plan for institutional strengthening, most large cities should establish clearly defined jurisdictional arrangements that take into account the cross-media and cross-sectoral nature of many urban environmental concerns.

Building local capacity

3.17 One of the keys to effective environmental management is to develop the institutional and technical capacities of local agencies responsible for environmental infrastructure and urban

Box 3.2. Community provision of low-cost sanitation

Community-based Management in Northeast Brazil

In the cities of northeastern Brazil, communities have been managing condominial sewerage systems that connect inexpensively to a block of houses. Success depends on residents jointly allowing the systems to be built on their land—thus the term condominial. Instead of digging under the streets in front of the houses (an expensive undertaking), a short grid of small, shallow "feeder" sewers are run though backyards. These innovations cut construction costs by 20-30 percent. Residents choose their level of service, and are responsible for operating and maintaining the feeder. Residents can choose: (i) to continue with their current sanitation system; (ii) to connect to a conventional water-borne system; or (iii) to connect to a condominial system. Families are free to continue with their current system (which usually means a holding tank discharging into an open street drain). In most cases, however, those families who initially choose not to connect eventually end up connecting. Either they succumb to heavy pressure from their neighbors or find the build-up of wastewater in and around their houses intolerable once the connected neighbors fill in the rest of the open drain.

The Orangi Project: Innovative Sewerage in Karachi

A small amount of core external funding started the Orangi Pilot Project (OPP) in 1980. The purpose of the OPP was to promote community self-help and the provision of affordable sewerage systems in Karachi's squatter settlements and to develop organizations that could provide and operate the systems. Coupled with an elimination of corruption and the provision of labor by community members, the costs (in-house sanitary latrine and house sewer on the plot, and underground sewers in the lanes and streets) are less than $70 a household. The OPP staff has played a catalytic role by explaining the benefits of sanitation and the technical possibilities to residents and by conducting research and providing technical assistance. The OPP staff never handled the community's money. The household's responsibilities included financing their share of the costs, particularly in construction, and electing a "lane manager," who typically represents about fifteen households. The lane committees, in turn, elect members of neighborhood committees (typically around 600 houses) who manage the secondary sewers. Although the OPP concept includes municipal government subsidy and later cost recovery for sewer trunks and treatment plants, the state is financing these facilities without cost recovery. The OPP's early successes created a "snowball" effect, in part because of increases in the value of property where lands had been installed. As the power of the OPP-related organizations increased, they were able to bring pressure on the municipality to provide municipal funds for the construction of primary and secondary sewers. Sewerage has now been provided to more than 600,000 poor people in Karachi and several progressive municipal development authorities in Pakistan have attempted to follow the OPP method. The OPP approach of community involvement and responsibility has been adapted in Cairo, Egypt, where the Zabbaleen community, with support from the Ford Foundation and the World Bank, have developed a successful waste recycling enterprise that has spun off into several entrepreneurial efforts.

The OPP has evolved into the OPP-Research and Training Institute and has developed model programs for low-cost housing, for basic health and family planning, for job- and entrepreneur-training for women, for supervised credit for microenterprises, and for upgrading privately-run schools.

Sources: northeast Brazil: World Bank (1992); Karachi: Kahn (1992)

services as well as for formulating, implementing, and enforcing locally appropriate pollution control policies and controls. For example, adequate operational management of urban services will go along way toward effective environmental management through the introduction of the principles of strategic planning, accountability, efficiency, maintenance, feedback, and cost recovery. Similarly, ensuring effective enforcement capacity, which requires strengthening human and technical capacity as well as increased financial resources, is crucial to the success of any policy or regulation affecting urban environmental quality. For most cities, building capacity in urban environmental planning and management will involve diverse skills and capabilities, many public and private actors, and a range of tools for capacity-building.

3.18 ***Skills and capabilities.*** Managerial, regulatory, technical, and financial skills are required for effective environmental planning and management. Within each of these categories a wide range of capabilities are needed. Some of the required skills are as follows:

- Key managerial capabilities—policymaking, conflict resolution, establishing administrative and public participation processes, and developing training programs and information systems.

- Key technical capabilities—planning, operations, and maintenance in key areas of urban environmental management.

- Regulatory capabilities—revising laws and codes and setting regulatory standards.

- Financial capabilities—capital budgeting, municipal accounting, setting tariff and tax structures, and establishing revenue collection procedures and financial incentive programs.

Capacity building for key actors

3.19 The two groups of actors that have a significant influence on environmental problems include the (i) the public sector, which encompasses all relevant agencies, departments, and authorities at the national and local or metropolitan level as well as politicians at all levels; and (ii) the private sector, which includes formal industries, informal enterprises and services, and local communities and relevant NGOs. Capacity building for the constituents of each group should correspond to the environmental management role each participant is expected to play.

3.20 ***Public sector.*** Effective urban environmental management will require substantial capacity building at the national as well as local (or metropolitan) levels. National authorities should build regulatory capacity by revising or developing new legislation and policies. Such regulatory efforts should promote investments in environmental improvement, permit use of a full range of low-cost technologies for various management services, and encourage private sector participation in urban service delivery and in other aspects of urban environmental management. Environmental planning and management at the national level should focus on the following:

- selecting appropriate policy instruments to meet pollution control, waste management, and land management objectives in urban areas;

- establishing environmental standards and effective monitoring and enforcement programs;

- establishing and administering programs of technical and financial assistance to support environmental agencies at lower levels of government; and

- empowering local governments to carry out local environmental management, establishing standards for local performance, and adopting measures to ensure accountability.

Box 3.3. Urban environmental management in Tianjin

Tianjin, China's third largest urban area, has suffered from concentrated air and water pollutants, traffic congestion, environmental health problems, a deteriorating housing stock, and seismic damage. Over the past decade, the quality of several key environmental indicators improved during a period of rapid industrial growth and increase in the motorized vehicle fleet. Much of this has been due to building up urban environmental management capacity.

Since environmental protection activities in Tianjin began in the 1950s, forty-six governmental organizations at or above the county level have been created to focus on environmental problems in the municipality, mostly under the umbrella of the Tianjin Municipal Environmental Protection Bureau (TEPB). Management of the urban environment gained momentum when the different sectors of the community were mobilized to support environmental and other improvements, primarily post-earthquake reconstruction and infrastructure upgrading. This was largely achieved by raising financial resources from enterprises and government bureaus for large public works projects, by mobilizing volunteer workers, and by building the institutional and financial capacity for ongoing environmental management.

The TEPB now has a large staff divided into departments that cover planning and finance, law and policy, science and technology, water quality, air pollution, natural resource protection, monitoring, supervision of effluent charges and penalties, inspection, and education. Some staff work directly for the TEPB; others work for related agencies. Over 80 percent of TEPB's budget comes from locally-collected discharge fees; the remainder comes from fines, local taxes, central government grants, and contracts. The majority of its expenditure goes to investment in fixed assets for environmental protection (95% of expenses), with small amounts for maintenance, personnel, business activities, and instrumentation. Some of the innovative activities that Tianjin has pioneered over the past fifteen years include:

- **Integration of environment in urban planning.** Air, water and noise targets are established and enforced according to the ecological features, geographic characteristics and economic development activities of three municipal zones.
- **Use of regulatory and economic instruments.** Municipal officials use standards and permits, industrial zoning, monitoring of emissions and ambient conditions, environmental impact assessments, fiscal measures to encourage environmental improvements, and economic incentives.
- **Discharge fee and pollution control fund.** A noncompliance fee is collected from enterprises that emit air pollutants, waste water, and solid waste. Eighty percent of collected fees are placed in a fund that is returned to the enterprises in the form of concessional loans and grants to finance pollution control investments.
- **Direct investment in infrastructure.** The municipality has encouraged construction with positive environmental effects.
- **Slum upgrading and housing redevelopment.** The municipality has relocated houses and industrial facilities in order to develop land more in line with locational values and environmental considerations.

Source: Leitmann (1991)

3.21 Capacity building for local authorities should focus on establishing appropriate institutional arrangements and strengthening managerial, technical, and financial capabilities. To facilitate integrated planning and management, local authorities should be able to perform cross-sectoral environmental planning and management tasks efficiently and effectively. This will involve clarifying institutional arrangements and administrative routines; establishing inter-agency linkages, information flows, and consultative processes; setting up a local steering committee for environmental planning and management; and modifying or removing institutional structures or administrative procedures that cause bottlenecks or duplication of effort.

3.22 Ideally, environmental protection functions should be organized and coordinated across sectors, with responsibility delegated to the most appropriate level of local government, depending on the nature and scale of impact of the problem and requisite solutions. In some instances new environmental agencies may need to be established. In Tianjin, China, for example, many new local government organizations were created to address urban environmental problems (see Box 3.3).

3.23 Local authorities require a wide range of skills and capabilities in integrated environmental planning and urban service delivery. For example, local government officials will need to upgrade their skills in such areas as drafting legislation and regulations, monitoring and enforcement, project preparation and feasibility analysis, environmental impact assessment, policy making, administration, conflict resolution, and public education and outreach. These skills are needed in such sector-specific areas as solid waste management, hazardous waste management, air pollution control, disaster preparedness, water supply, sanitation, drainage, groundwater protection, land management, cultural resource protection, and coastal zone management. Capacity building in urban management and municipal finance is also essential. For example, municipalities will need to set up the right revenue structures to finance recurring costs and will need to obtain the necessary financing for capital investments. Capacity building in property tax administration also is vital if the property tax is used for customer billings and revenue collection as well as to finance local environmental services such as drainage and public cleansing. Finally, cost accounting systems are needed to control service and monitor for cost effectiveness and efficiency.

3.24 ***Private sector.*** Both the formal and informal private sector should build capacity in various aspects of environmental management. Formal industries, for example, need to strengthen pollution control technology and efficient service delivery. Capacity building also should extend to private research and development concerning industrial pollution control technology or other aspects of environmental management. In addition land developers also will need to develop the necessary skills to ensure environmentally safe land development and housing construction. Informal service enterprises must learn how to organize their operations to facilitate their assimilation into such formal urban operations as solid waste collection and disposal.

3.25 Similarly, NGOs, local interest groups, or other CBOs should be encouraged (through active support and capacity building) to participate in environmental planning and decisionmaking. Capacity building for this cast of actors should focus on public education, consensus-building, and organizing and mobilizing public participation. Those NGOs that are not familiar with the technical aspects of environmental issues could benefit from the assistance of specialized consultants or others with technical expertise in key environmental areas.

Tools for capacity building

3.26 To acquire the skills needed for effective environmental management, several capacity building tools should be developed and used. Public sector training and technical assistance can be expanded to meet environmental management goals. Private sector technology transfers can be encouraged through partnership arrangements, and public information and outreach programs can be designed to promote public participation and support.

3.27 ***Training.*** Training for local government professional and managerial staff should cover a broad range of topics such as institutional analysis, project preparation and feasibility analysis, selection of locally appropriate regulatory and economic instruments, and the mobilization of broad-based public participation. Training can be undertaken at all levels within the organizations responsible for infrastructure provision, water quality management, solid waste management, and land management. It can include both in-country (formal and in-service) and overseas managerial, technical, and engineering courses as well as training courses for laboratory and field technicians. Similarly, the formal and informal private sectors need training not only to build capacity for carrying out various pollution control or waste management operations, but to build awareness and understanding of the causes, complexities, and impacts of environmental degradation.

3.28 ***Technical assistance.*** As part of their decentralization programs, national governments can establish national technical assistance programs to strengthen municipal government. To supplement standard training programs, for example, government authorities as well as NGOs might consider setting up a dedicated institute for training and organizing a cadre of urban managers, administrators and technical staff, and other relevant actors to provide cities with the skills necessary to successfully manage environmentally sound urban development and growth. Representatives of such an institute could go from one urban area to another, helping to develop institutional frameworks, building analytic capability, installing monitoring and analytic software, and then return periodically to review results and provide further advice. Over time this could develop into a permanent extension service, such as IBAM (Brazilian Institute of Municipal Administration) in Brazil.

3.29 ***Private sector participation.*** The private sector can be a tool for transferring environmental technology and innovative management techniques. Participation can come through a joint venture between a foreign firm with a new technology to sell and a local firm wanting to use that technology. An example where this often occurs is in hazardous waste management where a foreign firm brings in its technology to help a local firm solve a local urban hazardous waste management problem. Ideally, in such circumstances the technology is truly transferred, so that local operators and managers can take over operations and master the technology themselves. Private sector participation is also encouraged in areas where government performance tends to be weak, as in solid waste management. In Latin America the privatization of solid waste management functions is now widespread; the phenomenon has allowed the introduction of new technologies (for collection and disposal) where none existed before. Another important vehicle for private sector involvement in technical assistance and capacity building is industry associations.

3.30 ***Public information and outreach programs.*** Effective public information and outreach programs are needed to promote public participation in environmental planning and management as well as to create incentives for improved environmental services. These programs, which can be carried out by local authorities, NGOs, or CBOs, should be directed to all sectors and levels of society. This means reaching local industries, NGOs, community organizations, vulnerable groups, and others who need to understand the consequences of their policies or actions and the range of available options for improvements. Particular attention should be directed at the education of women because of their prominent role in household decisionmaking related to waste-related practices as well as in educating the next generation. In Karachi, Pakistan, for example, a publicly sponsored squatter improvement program paid off by investing in women. A soakpit latrine pilot project in the low-income settlement of Baldia empowered women by teaching them that better sanitation (supplied through community labor and supervision) could improve family health and living conditions. Due to its initial success, the project progressed beyond building latrines to the provision of other community services (Water and Sanitation for Health Project 1993).

3.31 ***Delineating and defining institutional arrangements.*** Because environmental problems cut across political boundaries, institutional arrangements and responsibilities need to be clearly delineated. Typically, national environmental authorities are responsible for establishing environmental policies, regulations and standards. They tend to be responsible for providing technical and financial assistance to local governments for program implementation. State and local governments as well as special regional authorities (for example, air quality management districts, river basin commissions, metropolitan waste disposal authorities) take enforcement actions to control private polluters.

3.32 Municipal government is almost always responsible for domestic sewerage and solid waste collection and disposal, and in failing to manage these activities properly, city authorities are polluters. Therefore, a clear distinction must be drawn between the polluting activities of local government and those under local government control (for example, industrial emissions in urban areas, industrial discharges into municipal sewers, hazardous waste disposal in municipal landfills). Regional, state, or national authorities must be able to enforce municipal waste disposal regulations either through powers of persuasion or coercion.

3.33 To complement government responsibilities in the formulation and enforcement of pollution control policies and regulations, the role of community groups, environmental NGOs, and the private and informal sectors must be recognized and encouraged through self imposed legislation. The Thai Industrial Environment Fund for hazardous waste disposal is a good example. To minimize monitoring and enforcement costs while ensuring that all waste is collected and its disposal paid for by the responsible firm, Thai factories would be required to deposit their waste disposal fees for the entire year in advance with the Fund, along with a matching bond or guarantee. The bond would be progressively returned to the company when it delivered waste to the treatment facility. Environmental audits of polluting industries are another form of private sector regulation. These audits involve yearly assessment of the environmental impacts of a company's operations as well as its compliance with applicable standards and regulations. Private environmental management firms can be hired either by other private companies to carry out these audits or by government agencies that can integrate the audits into government monitoring and regulatory procedures as well as waste disposal and hazardous waste management programs (Kaji 1992).

Improving the operation of urban services

3.34 Emphasis should be placed on upgrading the coverage and management of urban infrastructure and services and increasing the efficiency and effectiveness of local investments in environmental infrastructure and services. Ensuring environmental protection often requires complementary urban services. For example, the introduction of piped water generally requires parallel investment in sanitation; improvements in solid waste collection should be accompanied by corresponding investment in safe disposal; investments in storm water drainage usually requires improved solid waste management. To promote efficiency in service provision, governments should consider sharing responsibilities and risks with the private sector. Public-private partnerships can help to secure needed financing for public services, reduce costs, increase efficiency and to acquire access to more appropriate equipment and services.

3.35 Effective management of urban environmental services is a pre-condition for improving urban environmental quality. Although a comprehensive approach to urban environmental management is required for maximum effectiveness and protection, each sectoral agency will still need to carry out its own planning, management, and operations.

3.36 ***Urban water supply and sanitation.*** The provision of clean water and sanitation is one of the most critical needs confronting developing country cities. Water supply management requires effective pollution control systems (for both surface and ground water supplies); pricing strategies that encourage both household and industrial conservation; and, where appropriate, targeted subsidies, including cross-subsidies to provide expanded coverage to low-income populations. Above all, efficient, flexible approaches are needed. For example, demand-based sanitation planning recognizes that some parts of the city will be serviced by conventional sewerage (with an average household investment ranging from $300-$1,000), while other parts will be serviced by lower cost systems. Potential alternatives to conventional sewers include pour-flush latrines (typically about $100-$200 per household); simplified or small-bore sewers (about 30 percent less than conventional sewerage); and the condominial system, as in the example of northeast Brazil. A key focus in efforts to provide water and sanitation services, therefore, should be on community ability and willingness-to-pay for specific types and levels of service.

3.37 ***Storm water drainage.*** Storm water drainage should be an integral component of any wastewater disposal or pollution control system. For low-income areas, low-cost participatory approaches often are the most appropriate. For maximum cost-effectiveness and efficiency, the most appropriate institutional arrangement involves NGO or community self-help efforts with government assistance in the form of technical supervision or assistance. Due care is required to ensure that local drains tie into the city-wide drainage systems. In Surabaya, Indonesia, for example, local communities recently insisted that, before they make improvements to tertiary drains, secondary and major drainage systems must be upgraded. They understand an important priority: in several flood-prone areas near the outskirts of the city, so-called drainage channels are in fact irrigation canals originally designed to bring water in, and not out. Until major investments are made to deepen and widen the drains, the problem will persist (World Bank 1993). Such drainage improvements need to be coordinated with road construction and land use planning to prevent new developments from blocking off natural drainage patterns. Also, in all cities, drainage services need to be coordinated with the municipal solid waste management system to ensure that uncollected waste will not end up in open drains and cause flooding and health problems.

3.38 ***Municipal solid waste management.*** Municipal solid waste management in most cities needs urgent attention. Future improvements should incorporate a comprehensive policy framework and strategic solid waste management plan that take into account all physical, technical, legal, institutional, financial, environmental, and socio-cultural aspects of solid waste management. In most cities, institutional strengthening and sound management practices are needed to achieve expanded coverage for low-income groups, efficient service provision, and environmentally sound disposal operations. In large metropolitan areas collection will best be decentralized to the municipal level or lower. Disposal responsibilities, however, should probably be centralized at the metropolitan level (sometimes covering several municipal jurisdictions). To provide more efficient services, the potential role of the private sector (both formal and informal) should be explored and exploited further. In addition, more attention needs to be directed to financing capital investments and to covering recurrent costs of the solid waste service. Municipalities should attempt to recover the maximum amount of collection costs through user charges. Recurrent disposal costs may be recovered collectively through local taxes and intergovernmental transfers. To raise the quality of solid waste services, however, most investments will need to be accompanied by broader improvements in municipal finance, accounting, and budgetary control (Bartone, Bernstein and Wright 1990).

3.39 ***Urban transport management.*** Urban authorities have an important role to play in managing urban transport systems and therefore emissions from motor vehicles, road congestion, and road accidents. Local governments often make decisions about public transport options and transport modes (choosing the right blend of private- and public-transport services and the mixture of vehicles powered by alternative fuels, electric trolleys, subways, bicycles, and so forth). Road construction and maintenance, traffic management, vehicle inspection, road or vehicle use restrictions, and congestion pricing are also handled by local governments in whole or in part. Apart from improving operational efficiency, urban transport authorities can promote efficient urban land use by providing public transport to service high density areas. A transport-efficient land use pattern may include high-density, mixed-use development in central urban areas with several outlying high-density areas—all of which can be linked by an extensive public transport system. This approach has been adopted in Curitiba, Brazil where careful integrated transport and land use planning helped shape the city into a pleasant, low-congestion place to live and work (see Box 3.4). More often than

Box 3.4. Integrated planning and environmental improvements in Curitiba, Brazil

By the late 1960s, Curitiba, with a metropolitan population of 800,000 people (now over 2.2 million), was becoming another automobile-dominated city plagued by traffic congestion, urban sprawl, and inadequate public transportation. The municipal government initiated a series of integrated measures to encourage higher densities around major public transportation routes, reduce dependency on the private automobile, and make the city more accessible to pedestrians and cyclists. These measures have been sustained and improved upon for over twenty years and have produced impressive results:

Measures	Effects
Integration of transport and land use policies: Land use legislation enforces higher densities around major transport corridors and roads are geared to land use in each area.	Curitiba has one of the lowest rates of ambient air pollution in Brazil; fuel consumption has been reduced by 25 percent, with gasoline use per vehicle 30 percent less than in other Brazilian cities.
Public transport: Main roads have an express bus lane; different lines are integrated for rapid transfers; the system is faster and cheaper than those in other Brazilian cities.	The bus system serves 1.3 million passengers daily, or 75 percent of all commuters; people spend about 10 percent of their income on transport, one of the lowest rates in Brazil.
Provisions for pedestrians and cyclists: A network of bike paths links city neighborhoods and parks; pedestrians have priority in downtown areas with streets closed off to form pedestrian malls.	Once-declining shopping districts are now lively and profitable; there is little traffic congestion although Curitiba has over 500,000 cars (second highest amount per capita in Brazil).
Preservation policies: New development is concentrated in existing urban space; old structures are renovated for new uses.	Historic buildings are preserved and degraded land is reclaimed for such uses as an opera and a botanical garden; green space per capita has expanded from 0.5 to 50 m^3.
Traffic control: Some streets are closed to autos; in others, speed limits and trees slow car traffic.	Curitiba's rate of accidents per vehicle is now the lowest in Brazil.

Source: Rabinovitch w. Leitmann (1993); Lowe (1992)

not, cities must develop approaches together with the national authorities that normally determine fuel options (such as lead-free gas) or mandatory vehicle emission control technologies, as in Mexico City (see Case Study 6 in the Annex).

3.40 ***Urban land management.*** Effective urban land management is crucial for mitigating the impacts of urbanization on environmentally fragile land and other resources. Land management can also help reduce the vulnerability of low-income groups to both natural and man-made hazards and can play a role in mitigating transport-related air pollution. For rapidly expanding cities, a key challenge is to reach a proper balance between urban development and environmental protection. This will require a two-part urban land management strategy designed to enable land markets to function and to protect priority land and cultural resources. Part one involves providing infrastructure so as to reduce risks to developers and to encourage high density development where desirable. Part two involves formulating regulations that are appropriate to local conditions, establishing flexible development standards, and minimizing administrative requirements.

3.41 Measures to protect priority resources include:

- Restricting development in specific areas through special controls, development standards, or taxes—coupled with effective enforcement, and when necessary, provision of alternative locations.
- Allowing or encouraging appropriate use of sensitive land or cultural property accompanied by the necessary impact mitigation measures and management controls.
- Encouraging preservation through tax incentives.
- Improving city-wide environmental management to limit adverse externalities.

3.42 Management tools for carrying out these strategies include locally appropriate property tax systems; land use planning and controls; land registration and information systems; and land acquisition. Once these tools are in place, consideration can be given to regularizing land tenure and providing infrastructure and housing improvements. In Curitiba, for example, the coordination of land use controls and transport investments has reduced the negative environmental effects of an automobile-intensive transport system. In Benin, where the capital coastal city of Cotonou is subject to serious flooding because of its location and because of the neglect of road and drain maintenance, the government is investing in extensive citywide improvements to infrastructure, urban services, and municipal finance. No single instrument will be effective in achieving all of the city's land management objectives. Urban land managers will need to select the most appropriate instrument or mix of instruments to meet the particular needs and responsibilities of the various public and private actors in the land market.

3.43 ***Other urban functions.*** Health care and education are two environmentally important activities often managed at the state, metropolitan, or municipal level. The decentralization of health services to local communities can promote greater community involvement, contain costs, and reduce duplication of services. And decentralization can facilitate a more rational and unified health service on the basis of geographic and administrative areas (such as by district) and encourage increased community financing and control over primary health care facilities and staff (Vaughn 1990). On the education front, if cities are responsible for their school districts, they can also introduce urban environmental issues into their curricula.

Establishing public-private partnerships

3.44 Several benefits can arise out of effective partnerships between the public and the private sector. Private companies, informal sector enterprises, CBOs, and NGOs can provide urban environmental services, mobilize financing (or voluntary labor), introduce technological innovation and adaption, and develop urban land. Public agencies can facilitate this by ensuring a supportive legal and institutional setting for private sector involvement. This can mean setting prices at their true economic costs and ensuring that risk is shared equitably among actors involved in service delivery and other activities. The private sector in turn must accept the basic premise that improving environmental quality is good for business—the environmental sector is a "growth" industry and pollution control and clean-up present potential new market niches for private industry.

3.45 Based on various studies, the immediate result of pollution control legislation is new investments in clean technologies and the corresponding creation of new jobs. In the United States, for example, Dupont forecasts annual revenues of $1 billion by the year 2000 from its safety and environmental resources division set up to help industrial customers clean up toxic wastes. Further, the modernization of basic environmental infrastructure (for example, water and wastewater treatment plants and waste disposal facilities) creates a variety of highly skilled jobs. Clean-up and conservation activities are highly labor-intensive and therefore an important source of job creation. By contrast, failure to take environmental considerations into account can actually have a devastating impact on employment. For example, several North American chemical plants have faced the threat of closure because they have polluted local water supplies (Megalli 1991).

3.46 ***Delivery of environmental services.*** Where market and regulatory conditions are favorable, public/private partnerships can provide another avenue for environmental capacity building. In the case of municipal solid waste management, private companies often can operate more efficiently than the public sector, as long as there are exclusive service districts and competitive bidding. To maximize the effectiveness and efficiency of private sector services, however, it also will be necessary to strengthen local public capacity in the areas of contract preparation and supervision. To promote public-private partnerships in urban environmental services, developing countries should establish enabling legislation and have well-functioning markets and a clear understanding of what populations are willing and able to pay for improved services. Public-private partnerships can be between a municipality, an NGO, and the community; or between the city and the informal sector.

3.47 In many instances, the private sector is better suited than the government to deliver public services where competition and accountability to regulators are present. In many developing countries, for example, the private sector handles solid waste collection and disposal operations; the local government's role is to create competitive conditions, write performance-based specifications and contracts, monitor and regulate contracts, and protect public interests. In Latin America, studies of private sector provision of municipal solid waste services in Buenos Aires, Caracas, Santiago, and São Paulo confirm that, where firms have exclusive rights to service-specific areas and where contracts are awarded by competitive bidding, private firms can deliver services, contain costs, and maintain high quality service (Bartone et al. 1991). Regulatory systems, therefore, should enable private companies to supply urban services as long as the local government can develop adequate performance specifications, ensure competitive bidding, tie payment to performance, and oversee contract performance (Cointreau-Levine 1994).

Box 3.5. Private sector service delivery

The Agence d'Execution des Travaux d'Interêt Publique (AGETIP) is a private, not-for-profit legal enterprise that contracts with Senegal's central government to execute urban infrastructure works and urban service projects (similar agencies operate in Niger, Benin, Burkina Faso, Mali, and Mauritania and others are being formed in the Gambia, Chad, Madagascar, and Togo). The arrangement is stipulated in several documents including a manual defining the duties and responsibilities of the two parties. Municipal and central governments sign specific delegated contract management agreements with AGETIP each time they submit a sub-project for execution. AGETIP hires consultants to prepare designs and bidding documents and to supervise works; issues calls for bids, evaluates and adjudicates the bids, and signs the contracts; and evaluates progress, pays the contractors, and represents the owner at the final hand-over of the works. As of January 1993, Senegal's AGETIP had implemented 330 projects for a total amount of $55 million. These projects are located in seventy eight municipalities and have created 50,600 temporary jobs and 1,500 permanent jobs. Over half of the projects executed have dealt with the environment (drainage, garbage collection, canal clearing, sidewalk improvements, and road maintenance). The AGETIP model creates jobs for low-income groups through its labor-intensive methods. The "contracting-out" approach creates demand for the services of local contracting and consulting industries, thus stimulating their development and increasing spending in local economies. Its open competitive bidding process that eliminates wasteful and inefficient operators, an effective management information system, and a sense of accountability that allows for timely decisionmaking—all of which inspire other agencies to improve their own performance.

Source: Pean (1993)

3.48 A second form of partnership is between a municipal government and an NGO that arranges private sector service delivery. In Dakar, Senegal, the creation of a not-for-profit contract management organization known as Agence d'Execution des Travaux d'Interêt Publique (AGETIP) allowed the public sector to manage the contracting of private services to undertake urban infrastructure and works on behalf of local communities. The efficiency of service delivery and the living conditions of the poor were improved. This approach is now being replicated in Burkina Faso, Mali, and Niger (see Box 3.5). A variation of this approach is for the NGO to provide environmental services directly to the community. In the case of the Orangi Pilot Project, an NGO has succeeded in providing latrines and sewerage to thousands of households in Karachi's largest slum.

3.49 The role of the informal sector in service provision can be equally significant. For example, local waste authorities may be able to legitimize and integrate the informal sector into the formal solid waste management system. An informal sector enterprise can be contracted to work on collection, resource recovery, and recycling. Experience in Cairo and Ciudad Juarez (Mexico) illustrate how informal groups have been successfully converted into recycling cooperatives and granted concessions for collection and recycling (see Box 3.6).

3.50 ***Technological innovation and adaption.*** Partnerships between governments and businesses can stimulate the development and adoption of affordable and environmentally sound technologies. For example, private enterprises and research institutes can develop clean technologies for energy production and manufacturing, promote new approaches to solving inner city transport problems to improve the productivity and quality of life of cities, and develop new resource conservation technologies for industry. The government can encourage technology transfer by cultivating contacts with international firms, by providing incentives for private sector participation (for example, research grants), by establishing research and development organizations to pursue promising new technologies, and by structuring tax laws to reward the retrofitting of older industrial stock (Kaji 1992). Take, for example, the introduction of unleaded fuel in Thailand: citizens

Box 3.6. Resource recovery by the informal sector

Cairo: In Cairo, an informal sector group of garbage collectors, known as Zabbaleen, and local contractors, known as Wahis, was transformed into the private Environmental Protection Company (EPC). EPC has the contract for waste collection in several parts of the city. Although the responsibilities for solid waste management have long been shared by the municipal sanitation service and the Zabbaleen, the formation of the EPC established the Wahis and Zabbaleen as key participants in the local governor's program to upgrade solid waste management in Cairo. The Wahis administer the system, market the company's services, collect household charges, and supervise service deliveries. The Zabbaleen collect and transport the waste, supplying their labor in exchange rights to recycle the waste. After establishing the EPC, the Wahis increased their earnings and the Zabbaleen earned additional income and were able to reduce collection time by 30-50 percent. The Cairo Governorate is seeking to extend the EPC service to other areas of the city.

Cuidad Juarez: In the northern part of Mexico, the establishment of a small-scale waste recovery cooperative has improved conditions for waste recovery workers living by the tip site of Ciudad Juarez. Since 1975, the Cooperative Society for Material Recoverers for Juarez City (SOCOSEMA) has operated a waste recovery system sanctioned by the local authorities. SOCOSEMA's concession contract allows the cooperative to recover and sell collectively all types of industrial, commercial, and domestic refuse that can be found at the tip site as well as to import recyclable materials from the United States. SOCOSEMA pays a monthly sum of 75,000 pesos to the Municipal Treasury as a concession tax. After forming SOCOSEMA, the degree of actual resource recovery is similar to that of any other type of recycling facility. For workers, the impacts are notable. The system provides a relatively adequate level of income, a stable working environment, medical and educational benefits, and a strong sense of unity among workers.

Sources: Cairo: Environmental Quality International (1988); Juarez: Dagh-Watson (1987)

concerned about the increasing health hazards of air pollution in Bangkok formed an action group, risk assessment studies documented the problem, the national government accelerated the phase-out of leaded fuel through policy measures, and private sector petroleum companies quickly complied with government directives to bring cleaner fuels to the public (U.S. Agency for International Development 1992).

3.51 ***Land development.*** Public-private partnerships in land development can achieve environmental objectives and increase the supply of land for low-income housing. For example, under a land readjustment scheme, a public authority assembles privately owned land for conversion from rural to urban use, installs all public services, and finances the cost of the operation from the increase in land value resulting from the new infrastructure. By consolidating small property holdings and improving access and infrastructure, land readjustment can help ease development pressure on sensitive land that is environmentally unsuitable for urban development. Land readjustment also increases the efficiency of existing urban services and reduces costs for infrastructure provision. Such an approach has been effective in Indonesia, Korea, and Nepal. To protect sensitive lands, it also may be possible for the government to grant concession contracts to the private sector. For example, as part of the Guarapiranga Reservoir water catchment project being implemented in São Paulo (see Case Study 5 in the Annex), water quality will be protected by providing sewerage to the majority of illegal settlements and by creating buffer zones around sensitive parts of the reservoir to restrict further settlements. The government will also provide concession contracts to the private sector to manage the zones as public recreation areas. Similarly, public-private partnerships can be used to manage and protect cultural property, as is the case with the Historic Area Development Corporation proposed for Cairo (Serageldin 1992).

Choosing Appropriate Policy Instruments

3.52 Environmental authorities will need to select the most appropriate policy instrument or group of instruments to meet the particular needs, priorities, and special characteristics of each problem and locality. Instruments can include regulations, such economic tools as taxes or user fees, property rights, and land management instruments. Other instruments include tools to promote information and education. To select the most appropriate tools, and time(s) to intervene, the following factors should be considered:

- The desire for economic efficiency.
- The speed with which the environmental goals need to be attained (which determines the stringency of the reforms).
- Compatibility with existing administrative, political, and judicial frameworks, and with economic conditions, and the tax structure.
- The political acceptability of policy instrument.
- The complexity of application (for example, the method of computing charges).
- The ease of monitoring and enforcement.
- The policy measure's consistency with the overall environmental strategy.

Regulatory instruments

3.53 Strong regulatory policies are needed to mitigate or prevent serious degradation of water, air, and land resources. To be effective these policies must stress the development of appropriate standards and effective monitoring and enforcement systems. Such a framework could, for example, include the command-and-control approach to environmental management. This approach involves direct regulation, along with monitoring and enforcement systems; it relies primarily on the application of such regulatory instruments as discharge standards, permits and licenses, land and water use controls, and public health codes. Although this approach has been criticized for being economically inefficient and costly to enforce, command and control strategies afford a reasonable degree of predictability about how much pollution levels will be reduced and, when properly implemented and enforced, have gone a long way toward meeting environmental goals. Locally appropriate land use regulations are also needed to manage urban growth so as to minimize or avoid adverse environmental effects.

3.54 While strategic deregulation is sought in some areas (for example, relaxing land development standards so as to increase the supply of affordable land in environmentally sound locations), stricter regulation and enforcement are required in other areas (such as in controlling industrial emissions). In most countries, however, sustained regulatory actions are needed to meet long-term objectives. This will require the introduction of phased standards in line with financial capacity. Further, environmental legislation may need to be updated to incorporate new scientific knowledge, to phase in environmental regulations and standards in accordance with investment and enforcement capacity, and to establish new consultative mechanisms between different levels of government, business, and the public.

Economic instruments

3.55 When implemented well, economic (or market-based) instruments can be powerful tools for modifying the behavior of public and private polluters and resource users. They include user charges, resource pricing, pollution taxes, congestion charges, investment financing, construction grants, tax credits, and subsidies. Although some of these economic instruments involve the application of direct costs to the polluter (for example, charges based on the volume and toxicity of discharges, pay-per-bag systems for solid waste disposal, and permit fees for air emissions where the fee varies with the volume emitted), others will involve indirect costs, such as pollution taxes on inputs. Pricing industrial water use at its true marginal social and economic cost (of production, consumption, and treatment) can encourage companies to treat and conserve water. In Thailand, for example, many hotels along the country's eastern seaboard are treating and recycling their water for landscape irrigation because the cost of fresh water now exceeds the cost of treatment (Foster 1992).

3.56 Such incentive-based economic instruments as pollution charges, tax incentives, and targeted subsidies have been used in several countries to control water and air pollution. To various degrees, economic instruments incorporate the "polluter pays principle." When government forces prevail (rather than market forces), the polluter typically bears the cost of mitigation or prevention measures. The polluter's burden is determined by public authorities, based on what they feel is required to ensure environmental sustainability. In France, for example, effluent charges are levied on all public and private actors that pollute fresh or sea water. The resulting revenues are used to support water pollution control activities. Differential taxation or pricing can achieve similar effects. As another example, a special tax on leaded fuel can encourage the use of non-leaded fuel in automobiles while raising revenues for environmental management. In Thailand, pricing unleaded gasoline below the price of leaded gasoline prompted a switch to unleaded gasoline in less than five months.

3.57 Although economic incentives can reduce excessive reliance on regulation and investment programs to control pollution, they normally are not used alone to control pollution. In practice, they supplement direct environmental regulations and help raise revenues for financing pollution control activities or other environmental measures. Such incentives also encourage better enforcement and stimulate technical innovation. With water effluent control systems, for example, effluent fees and taxes have not eliminated the need to regulate discharges through permits (Bernstein 1991).

Property rights

3.58 The definition of property rights can be a powerful instrument for the environmental management of air, land, and water resources. Defining them can mean clarifying water rights to promote water conservation, allocating discharge or emissions rights to control pollution, and providing secure land tenure to encourage investment in housing and infrastructure improvements. The clear legal definition of water use rights has allowed the development of water markets in Chile. By contrast, uncertainty about water rights in Turkey caused the supreme court to strike down efforts by the city of Izmir to charge ground water extraction fees to private well owners.

3.59 Effective land tenure reforms can improve the efficiency of urban land development, encourage sound construction, and prevent disasters. The government can establish a system for clarifying land ownership and boundaries and can provide secure tenure to land occupants in illegal settlements, including those in hazard-prone areas. This will enable landowners to gain access to formal credit they need for investing in housing construction or improvements that will withstand certain hazards. Obtaining clear land tenure also may encourage households to demand rights to facilities and services that can improve health conditions in vulnerable areas. In Solo, Indonesia, for example, the opportunity to own land helped motivate slum dwellers to upgrade their plots and neighborhoods, resulting in major improvements to water supply, drainage, sanitation, and solid waste management (U.S. Agency for International Development 1992). Establishing such a system, however, normally requires the political will to institute land reforms at the national level. It also requires a review by experts of existing laws, the integration of customary systems of land tenure with modern land titling systems, and the design of simple and efficient titling and registration procedures. The provision of land tenure also may require incentives to encourage low-income landowners to register their properties and disincentives to ensure that the provision of tenure does not encourage further land invasions.

Land management instruments

3.60 Used judiciously, a range of land management instruments can help urban managers meet environmental objectives. These instruments include fee simple land acquisition (purchase of full title to land and all rights associated with that land) and expropriation (government acquisition of private land for public purposes upon just compensation), as well as less-than-fee acquisition tools such as easements, land exchanges, and the purchase of development rights. Other instruments may include donations and bargain sales, exaction (dedication), and land readjustment and guided land development. In Thailand, for example, some cities (including Udorn Thani, Kamphaeng Phet, Phayao, Songkhla, and Khon Kaen) have built new parks and recreation facilities using land and money donated by private individuals and companies. In Tunisia, much of the inner city land in Tunis comprising the *medina*, or old city, has been deeded to the religious authorities by private citizens. In land readjustment, a public authority converts rural or peri-urban land and extends public services to the area. These improvements are financed by the increase in land value resulting from the new infrastructure. Land adjustment has been effective in Germany, Japan, Korea, and Taiwan and has been adopted in Indonesia and Nepal (Bernstein 1993; MacAuslan and Farvacque 1992).

Information and education

3.61 Information and education are crucial for mobilizing participation in environmental activities and for expanding knowledge about environmental and health conditions and about the effects of various types of development on fragile or hazard-prone land. Tools to promote information and education include geographic information systems (GIS), land information systems, various types of assessments (for example, risk assessments, environmental impact assessments, and land market assessments), public information programs, private sector promotion, and research and monitoring.

3.62 ***Geographic information systems.*** GIS can be used by planners at several levels as a tool for assessing the impact of natural events on existing and proposed development activities (Organization of American States 1990). At the national level, GIS can be used to categorize land by the extent of risk to natural hazards, allowing an estimation of whether and to what extent natural phenomena pose a significant danger. At the regional level, GIS can be applied to more detailed studies to identify development potential and hazard-related constraints of selected areas. At the local level, planners can use GIS to formulate projects at the pre-feasibility and feasibility levels to locate such essential facilities as ports and airports, hospitals, health centers, police stations, fire stations, schools, energy infrastructure, emergency management facilities, and telecommunications. Hazard-free areas suitable for urban expansion or resettlement can also be identified. In Colombia, the Puerto Bogota Department of Cundinamarca used a GIS to identify hazard-free urban areas suitable for the relocation of thirty-four families living under high landslide risk. GIS is also an important tool for the operational management of many important environmental services such as water distribution and sewer pipe networks.

3.63 ***Land information systems***. A properly functioning parcel-based land information system can provide many benefits to local governments and private landowners. For local governments, such a system facilitates urban planning, land administration, land management, infrastructure provision, property taxation, and environmental assessment. For individuals and private enterprises, a parcel-based system facilitates real estate transactions in the following ways:

- It lowers transfer costs and eliminates delays when boundaries and ownership must be established prior to a transfer.
- It reduces litigation and associated costs.
- It improves accessibility of land information and thus allows a geographically larger and more competitive land market.
- It increases the availability of loan financing for all types of development or improvements (by providing credit institutions with reliable records of the extent and security of ownership).

3.64 ***Environmental impact assessments***. An environmental impact assessment (EIA) is an informational tool for analyzing the potential environmental impacts of a proposed project and its alternatives prior to implementation. The EIA ensures that the proposed project or action is environmentally sound and that any environmental consequences can be recognized early and taken into account during the project or program design. EIAs normally cover baseline environmental conditions; potential environmental impacts (both direct and indirect); comparison of the environmental impacts of alternative investments, sites, technologies, and designs; preventive, mitigatory, and compensatory measures; environmental management and training requirements; and provisions for incorporating input from the general public as well as for monitoring project impacts.

Closing the knowledge gap

3.65 Information also lays the groundwork for pulling in the broad cast of actors whose participation is needed for an urban environmental strategy to succeed. Closing the gap in knowledge about the urban environment can help spur action where it is needed most—at the community, industry, and local government levels. Although information requirements will vary according to city-specific problems and will depend on local capacity to generate data and analysis, the following types of information should be available:

- *Characteristics of media-specific environmental problems*. An assessment of the condition of environmental media (air, surface water, ground water, and land) is needed to determine the costs of interventions and appropriate management activities.

- *Dynamics of environmental degradation.* Knowing the nature, extent, impacts, and processes of environmental degradation as well as the relationship between environmental quality and development are essential preconditions to obtaining the political consensus needed to launch an urban environmental plan. These dynamics involve how environmental quality changes over time, the impact of environmental factors on development, the consequences of specific types of development on the environment, and the relationship of special groups (especially women, children, the elderly, and the disabled) to environmental problems.

- *Magnitude and distribution of impacts.* To set priorities for action, it is essential to know how various types of environmental degradation affect human mortality and morbidity, economic productivity, and ecosystem stability. Equally important are the monetary and other costs of degradation and the costs and benefits of different solutions.

3.66 To obtain environmental information, both public and private actors should strengthen their ability to routinely collect, assess, use, and disseminate environmental data. This will require an effective urban environmental monitoring program. In developing such a program, authorities should identify a limited set of environmental and performance indicators and then put in place the technical capacity to carry out accurate measurements and analyses on a regular basis. As capacity permits, more refined indicators can be added. The authority responsible for overseeing the monitoring program must specify sampling procedures, properly calibrate and maintain equipment, and ensure quality control of laboratory analyses. A unified system of monitoring and standardized analytical procedures will make it possible to assess the quality of the urban environment and judge the effectiveness of interventions over space and time.

3.67 Research and analysis alone will not close the knowledge gap—professional education and information exchange are essential for effective environmental management. Developing countries should incorporate curricula on the scientific, technical, and managerial aspects of urban waste management and pollution control into formal education programs. For example, environmental engineering programs should focus on conventional as well as non-conventional water supply and sewerage technology and solid waste management. Environmental education should be an integral component of any public or private waste management or pollution control initiative. Policymakers at all levels of government must be informed about the implications of their decisions, actions, or inaction in regards to critical waste and pollution-related issues. Engineers working for private industry also should be informed about new resource-efficient and environmental protection technologies. Public information campaigns should be directed at schools, communities, user groups (such as water users in a particular area), and industry groups.

Mutually supportive packages

3.68 Specific instruments alone will provide only short-term solutions to environmental problems. To have a lasting impact these tools need to be used in combination with other complementary policies. In Tianjin, China, for example, the municipal government has mobilized many actors and used many instruments to push through a broad range of innovative environmental management activities. These include industrial zoning regulations, a discharge fee, and a pollution control fund that levies polluting enterprises. The resulting revenues go into the fund, which extends concessional loans to enterprises to finance pollution control investments. Using land management instruments, the city has relocated houses and industrial facilities to develop land more in line with locational values and environmental considerations.

3.69 Comprehensive environmental management strategies such as Tianjin's take many years to install. The best way to launch such an effort is to start with cost-effective measures to tackle urgent problems in the context of a broad set of long-term goals. The city of São Paulo, for example, decided to first tackle water quality deterioration in the Guarapiranga reservoir as the initial step in its overall environmental action plan for the Guarapiranga Basin. Instruments to deal with immediate problems include land use rationalization and control, slum improvements and housing relocation, integrated provision of basic services in presently unserved areas, and the development of recreational parks to protect critical and damaged areas (see Case Study 5 in the Annex).

3.70 To succeed, mutually supportive packages should be designed to reduced several basic constraints to sound environmental management. Broadly, the constraints can be removed by mobilizing public support and participation, by raising awareness, by building constituencies that include the urban poor as well as powerful elites, and by promoting the participation of NGOs and representatives from the informal sector.

IV. FORMULATING AN URBAN ENVIRONMENTAL MANAGEMENT STRATEGY

4.1 With myriad urban problems and many tools for dealing with them, each city will need a process for determining the most appropriate mix of actions and investments that respond to its environmental priorities. When confronting environmental problems, cities exhibit different degrees of awareness, political commitment, and capacity to mobilize resources. The strategic approach to urban environmental planning and management suggested here is based on enabling participation and building commitment. It has been tested in industrialized and developing-country cities and is a viable approach for cities working toward setting up local versions of the urban component of Agenda 21. The approach can involve several activities, each of which should emphasize strengthening local capacity:

- **informed consultation** in which rapid assessments are conducted, environmental issues are clarified, key actors are drawn in, political commitment is achieved, and priorities are set through an informed consultative process;

- the formulation of an integrated **urban environmental management strategy** that embodies long-term goals and phased targets for meeting the goals; and agreement on **issues-oriented strategies** (that cut across the concerns of various actors) **and actor-specific action plans** (that cut across various issues) for achieving the targets, including the identification of least-cost project options, policy reforms, and institutional actions; and

- **follow-up and consolidation** in which agreed programs and projects are initiated, policy reforms and institutional arrangements are solidified, the overall process is made routine, and monitoring and evaluation procedures are put in place.

Urban Environmental Planning and Management

4.2 Urban environmental planning and management aims to identify urban environmental issues (ideally before they turn into costly emergencies), to agree on strategies and actions to resolve these issues among all those whose cooperation is required, and to implement these strategies through coordinated public and private actions. Over time the process should improve health and productivity in cities, reduce environmental hazards, and protect natural resources so as to sustain economic and social development.

4.3 Although urban environmental planning and management will differ from city to city, all urban areas can benefit by including eight basic elements in the process they adopt. These elements, also illustrated in Figure 4.1, are as follows:

- clarifying environmental issues to be addressed;
- involving those whose cooperation is required;
- setting priorities;
- negotiating issue-specific environmental management strategies;
- formulating an urban-wide environmental management strategy;

- agreeing on environmental action plans;
- initiating priority programs and projects;
- monitoring and evaluating progress and making periodic adjustments; and
- strengthening environmental planning and management capacity.

4.4 This approach, which has evolved from the experiences of many cities in both industrialized and developing countries, is being applied through such collaborative programs as the Urban Management Programme (SCP), the Metropolitan Environmental Improvement Program, and the Sustainable Cities Programme. These efforts, which center around helping cities initiate environmental management strategies, are described in detail later in this chapter (Box 4.6). Case studies of some individual city examples are described in the annexes.

4.5 The element that carries throughout is the need to strengthen environmental planning and management. Such institutional capacities as effective organizational arrangements, clear rules and procedures, trained professional and support staff, adequate facilities and equipment, and sufficient resources to cover recurring costs are critical prerequisites for successful environmental planning and management. Ideally these capacities should be "system-wide"—public and private sector actors should pool their skills to make the environmental management process. The efforts of all the actors should be bolstered by sound environmental analysis, which forms the basis for ranking problems and setting priority actions.

Figure 4.1. Sustainable Cities Programme environmental planning and management process

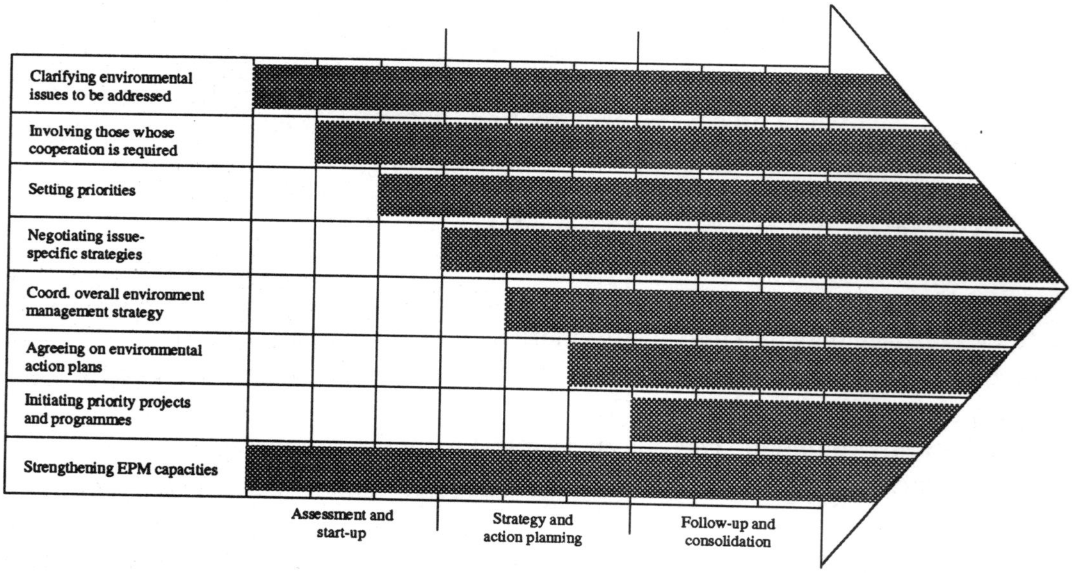

Source: Adapted from Jochen Eigen, 1992

Analysis versus process—a false dichotomy

4.6 Analysis and process go hand-in-hand in designing and implementing an environmental planning and management strategy. Many of the decisions made in early consultations, in setting priorities, and in project selection are based on good analysis. While the focus in this chapter is on process—the basic steps needed to carry out environmental management plans—analysis should not be given short shrift. Indeed, it is needed at most of the steps along the way, beginning with the initial rapid appraisal of a city's environment (based on existing data). The rapid appraisal should be conducted within an overall analytical framework. This framework guides early discussions and influences how environmental problems are ranked and how local priorities are set.

4.7 A balance must be struck between the findings of sound analytical studies (for example, cost-benefit analysis and risk assessment) and the concerns of various constituencies demanding actions that may not necessarily reflect the findings and recommendations of expert studies. And it should incorporate an integrated approach (that is, cross-media and cross-sectoral) to focus on analyzing and solving environmental problems while recognizing the need to designate specific agencies to be responsible for carrying out sector-specific improvements in urban service delivery and pollution control.

4.8 ***Getting the substance right.*** Getting the substance right means honing the analysis to answer urgent questions about priorities. It also means successfully translating analytical material and conveying it in a clear, compelling way to all the actors involved. Two good examples of where the substance was captured well are the Delaware Estuary and the Metropolitan Environmental Improvement Program projects (Case Study 4 in the Annex and Box 4.6, respectively). In the Delaware example, a Technical Advisory Committee monitored all results of a major research study and advised other committees on technical matters, including the Water Users Advisory Committee. Serving as a technical check on the planning process, the Technology Advisory Committee helped set water use and quality objectives. Involving all the actors in setting these objectives helped elucidate the cause/effect mechanisms relating waste load discharges to water quality and quantifying the costs and benefits associated with different control strategies. Much effort went into disseminating the results of the committee's research to other participants. Under the Metropolitan Environmental Improvement Program, work is under way in Bombay, Beijing, Colombo, Jakarta, and Manila to strengthen the capacity of pollution control and environmental protection agencies, especially in working with powerful economic planning and sectoral agencies at the local and national levels. This requires a thorough review of their efficiency and utility.

4.9 ***Setting priorities.*** Establishing priorities for action on the many environmental issues that may arise and the range of possible responses is not a simple task. The concerned or affected parties sometimes have conflicting agendas; human and financial resources may be scarce; objective measurement and analysis are difficult; and several diverse objectives must often be pursued simultaneously.

4.10 The way forward is provided through consultation—once local environmental issues have been identified and key stakeholders agree to participate, then it is possible to make an initial ranking of the top priority issues. Issues are ranked according to the impacts associated with each environmental problem and the local capacity to respond. Priority problems can then be selected on the basis of criteria such as:

- the magnitude of health impacts associated with the problem;
- the size of urban productivity losses caused by the problem;
- the relative impact of the problem borne by the urban poor;
- the degree to which the problem results in or is caused by unsustainable consumption of resources; and
- whether or not the problem leads to an irreversible outcome.

4.11 Although there is normally insufficient information to attempt rigorous analysis at this early stage, a conceptual cost-benefit framework is still useful for orienting the discussion. Current and projected trends in investment capacity can be compared with estimates of the damage that will occur without environmental improvements and the costs of cleaning up that damage.

4.12 As part of the consultative process, and with the support of technical specialists, the issues and possible responses can be tackled by local stakeholders. Priorities should be established systematically, starting with specific issues to be addressed, and identifying technical and operational responses, associated institutional requirements, and needed political support. The process should build awareness and a sense of ownership among stakeholders, and if successful, will lead to a consensus and political decision to proceed with the next steps. For example, participating in the Sustainable Cities Programme, the Tanzanian national government, municipal officials and private sector actors completed a five-day consultation during which they laid the groundwork for a city-wide environmental management strategy (see Case Study 1 in the Annex). To ensure continuity and prompt follow-up, a useful tactic is to tie the rapid assessment and consultation process directly to ongoing national- and externally-supported project identification or preparation efforts, as was the case in Durango, Mexico (Moguel and Velazquez 1992). Subsequent planning phases can then be included in formal project preparation activities and fed directly into investment decisions.

Ownership through consultation

4.13 The most effective way to build political commitment is to convince the actors that their involvement in the environmental planning process will pay off. They are more likely to feel a sense of ownership of the reforms if they are consulted at every important milestone. In conducting a rapid assessment for the metropolitan area of São Paulo, for example, local consultants prepared a profile and interviewed city, community, and business leaders to get their feedback. At a town meeting that included all the major actors involved, it was decided that cleaning up and protecting the city's Guarapiranga watershed was the most urgent priority (see Case Study 5 in the Annex).

A flexible approach

4.14 For any particular city, the starting point for environmental planning and management will depend on the general level of awareness about environmental problems, the degree of political will that exists to tackle priority issues, and the resources that can be mobilized to deal with priority problems. For example, Dar es Salaam and the secondary cities in Côte d'Ivoire described in the annex have limited resources and an emerging awareness and political commitment. In these cities the entry point was to carry out initial rapid appraisals and consultations with all interested actors in order to agree on priorities and build sufficient political commitment to adequately address them. In São Paulo and Mexico City, by contrast, a relatively high level of resources are available and the level of awareness and political commitment are strong and well-developed. There, the key actors involved have agreed on strategies for tackling priority problems and are carrying out actor-specific action plans and investment programs (for air pollution in Mexico City and for protection of the Guarapiranga reservoir in São Paulo).

Start Up and Assessment

4.15 The first step in environmental planning and management is to organize a forum and a planning process to clarify environmental issues that need to be addressed. A key goal here is to involve those whose cooperation is required and to establish their agreement on urban environmental priorities and procedures to be followed, and their political commitment to act on the priorities. Such political commitment often arises from a broad-based public concern about specific environmental issues. Although the impetus can come from any of the interested parties, such as a local environmental activist group, a local political leader or urban authority will have to move the process forward. Start-up and assessment typically results in the collection of existing city-level data and the preparation of an urban environmental profile taken from existing information. This phase also results in a carefully organized city consultation with the participation of all key actors to clarify roles, set priorities, establish procedures, and develop a sense of common purpose and ownership.

Clarifying environmental issues to be addressed

4.16 A rapid assessment of the urban environment is useful for screening those environmental issues that:

- harm the health and quality of life of urban dwellers;
- reduce the efficiency and productivity of cities;
- threaten permanent damage to local or regional ecosystems;
- cut across conventional lines of authority or responsibility, geographical boundaries, or time horizons; and
- require coordinated responses.

Initially, this screening process is achieved through a rapid assessment. First, researchers assemble and analyze such readily available information as: data from routine environmental quality monitoring, where they are available; data on existing environmental infrastructure and services; demographic, epidemiological and socioeconomic data; information on existing environmental policies, regulations, and institutions; and data on the natural ecosystems within or surrounding the urban area. To facilitate these assessments, several international groups have developed or are working on tools for urban environmental auditing. For example, the Urban Management Programme produced the urban environmental data questionnaire, which is a framework to guide the organization and collection of existing data (Box 4.1). In addition, several analytical tools, such as geographic information systems, are available to help interpret the information gathered in a rapid assessment exercise.

4.17 The next step involves assembling and analyzing available data and descriptive information and preparing an urban environmental profile. A generic outline for such a profile is provided in Box 4.2. Issues and institutional arrangements described in the profile will serve as the initial basis for determining actors whose cooperation is required. The profile provides a starting point for dialogue among stakeholders and helps identify information gaps. The UMP and the SCP have prepared and are in the process of preparing assessements and environmental profiles for Accra, Bangkok, Concepcion, Dar es Salaam, Guayaquil, Ibadan, Ismailia, Jakarta, Katowice, Madras, São Paulo, Tianjin, and Tunis. Single profiles covering groups of intermediate-size cities have been prepared for urban areas in the Singrauli region of India, in Côte d'Ivoire and in Mexico. A fuller description of the rapid assessment method is provided in a pair of UMP publications (Leitmann 1994).

Involving those whose cooperation is required

4.18 Urban environmental planning and management requires agreements and coordinated actions by a variety of public, private, and non-governmental actors at, individual, community, city

Box 4.1. Urban environmental data—outline of questionnaire

I. **Socioeconomic background conditions**
(Urban population, demographics, income and poverty, employment, municipal services, municipal expenditures)

II. **Housing conditions**
(Size, ownership, facilities, marginal units)

III. **Health conditions**
(Basic statistics, mortality rates)

IV. **Natural environment**
(Location, ecosystem type, meteorological data, dispersion conditions, topography, environmental hazards)

V. **Land use**
(Urban land use, newly incorporated urban land, land ownership, land registration, land use regulation, land market)

VI. **Urban transport**
(Basic statistics, vehicle stocks, motorized travel by mode, emissions, injuries from accidents, passenger car restrictions)

VII. **Energy use**
(Annual gross energy consumption, emissions from combustion, interconnected electricity grid, in-city electricity utility, urban electricity self-generation, household energy consumption, other indicators, energy pricing)

VIII. **Air pollution**
(Emissions intensity, emissions control, policy implementation, ambient concentrations, monitoring, environmental health)

IX. **Noise pollution**
(Noise levels, noise pollution control)

X. **Water and sanitation**
(Water resources, ground-water abstraction problems, future resources, water supply, water delivery, household sanitation installations, drainage network coverage, sewage flow rates, sewage treatment plants, sewage disposal, industrial effluents, water pollution policy instruments, water quality monitoring, monitoring)

XI. **Solid and hazardous wastes**
(Total solid wastes generated, municipal solid wastes, disposal of municipal solid wastes, municipal expenditures for solid waste management, dump sites, hazardous waste facilities, hazardous waste policies being implemented)

and national levels. As discussed in chapter 3, the planning process should involve those whose interests are affected by both environmental problems and proposed management strategies and action plans, those who control relevant instruments for implementing environmental management strategies, and those who possess relevant information and expertise needed for dealing with a wide spectrum of environmental issues. A list of potential stakeholders and participants is given in chapter 1.

4.19 Initially, the goal is to convince the stakeholders to make specific commitments to work on the environmental issues identified in the environmental profile. This is best achieved through a

Box 4.2. Generic outline for urban environmental program

I. Introduction
Background
Geophysical and Land Use
Socioeconomic Setting (demographics, economic structure, urban poverty)
City History: Environment-Development Linkages Over Time

II. Status of the environment in the urban region
Natural Resources
- Air Quality
- Water Quality (surface, ground, coastal, fisheries)
- Land (forests; agricultural land; parks and open spaces; historical and cultural sites)

Environmental Hazards
- Natural Risks
- Human-induced Risks

III Development-environment interactions
Water Supply
Sewerage and Sanitation
Flood Control
Solid Waste Management
Industrial Pollution Control/Hazardous Waste Management
Transportation and Telecommunications
Energy and Power Generation
Housing
Health Care
Rural-Urban Linkages
Other

IV. The setting for environmental management
Key Actors
- Government (central, regional, local)
- Private Sector
- Popular Sector (community groups and NGOs, media)

Management Functions
- Instruments of Intervention (legislative and regulatory; economic and fiscal; direct investment; planning and policy development; community organizations; education, training and research; promotion and protest)
- Environmental Coordination and Decisionmaking (mechanisms for public participation; inter- and cross-sectoral coordination; public-private linkages; information and technical expertise)

Constraints on Effective Management
Ongoing Initiatives for Institutional Strengthening

References

consultation process, typically starting with the collection of environmental and institutional information to be used in the profile, which is drafted and refined through a series of round table discussions or public fora. This culminates in an environmental consultation in which participants from the public, private, and non-governmental sectors consider environmental issues and possible responses at three levels of intervention—technical and operational, institutional and managerial, and political.

4.20 Efforts should be made at every stage of the planning process to define and delineate the roles of all key stakeholders. A steering committee should be established early in the consultation phase to help organize working groups, oversee the operations of the process, and ensure the required political support and resources. Eventually, working groups (made up of stakeholders and interdisciplinary technical support staff) can be formed around priority environmental issues.

The City-Specific Environmental Management Strategy

4.21 The next (and central stage) of the environmental planning and management process involves negotiating issue-specific strategies that respond to the city's unique mix of priorities and reconciling these strategies with an overall, coordinated urban environmental management strategy. This requires setting long-term environmental goals for the urban area or region, together with interim environmental objectives to guide phased investments and reforms. Key outputs include detailed information to clarify environmental problems and solutions, and issue-specific strategies agreed upon by all the key actors involved. The work is best done in issue-specific, inter-sectoral working groups.

Negotiating issue-specific environmental management strategies

4.22 In many countries planning is often short-term and incremental, as with the five-year development plans followed by many countries; in the public sector it is typically compartmentalized with little coordination across the conventional divisions of institutional responsibility. Coordination between the public and private sectors is minimal, as best. However, environmental management requires long-term strategic planning that is well-coordinated and that is endorsed by the many actors who shape urban development.

4.23 An issue-specific environmental management strategy will establish the framework within which short- and medium-term sectoral action planning and investments can occur. It should also provide a decisionmaking framework for private investments, recognizing that, over the long term, the investments will be made primarily by private actors (households, communities, and firms). The formulation of the strategy, therefore, should focus on achieving the coordination of both public and private decisions and actions.

4.24 ***Analyses for issue-specific environmental management strategies.*** To be effective, the formulation of issue-specific environmental management strategies must take into account the full range of required interventions. An effective approach is to intervene first at the technical level, where several instruments can be applied. Technical interventions allow fairly easy coordination among actors, and can be supported with information and technical expertise. A second intervention point is at the institutional level, where administrative and managerial capacities can be established, strengthened, and maintained. A third is at the political level, where institutional change must gain official support and adequate resources. The concerns of working group members at these operational, institutional, and political levels should be analyzed. Such an analysis can pinpoint the environmental issue early on, provide the basis for a work program, and structure a coordinated response.

4.25 To develop an issue-specific environmental management strategy, environmental problems and strategic interventions must be ranked. Typically this requires extensive negotiation and conflict resolution among the many actors involved. The issue-specific strategy that results from these activities can be based on several different kinds of analyses:

- *Health effects and environmental damages.* The first step in ranking environmental problems is to evaluate the health effects, productivity losses, ecosystem damages, and amenity losses associated with current and alternative levels of environmental degradation. If it proves too difficult to value impacts in economic terms, the analysis should at the least produce comparative risk assessments in terms of such quantitative measures as additional deaths, restricted activity days, units of production lost, or stocks of resources lost as a result of environmental damage due to specific problems. Specification of local cause-effect linkages is crucial, as is the evaluation of differential impacts on highly vulnerable groups such as the poor, women, children, and the elderly.

- *Regional economic efficiency.* This is an analysis of alternative long-term strategies for environmental management designed to achieve environmental quality goals and targets at the least cost to the urban region. Ideally, the cost-benefit framework could help rank priorities and be instrumental in the selection of the most appropriate level of environmental quality. Box 4.3 gives an example of how air pollution priorities were established in Tarnobrzeg, Poland using cost-benefit analysis. Where the benefits of environmental improvement cannot be accurately estimated, a simple least-cost analysis can be conducted. Target levels for environmental quality are then determined subjectively, using qualitative procedures for estimating reduced health impacts and environmental damages.

- *Financial feasibility.* This analysis concerns the capacity of the regional economy to finance the environmental protection investment needed to implement the long-term strategy and the ability of households, enterprises, and municipal governments in the urban region to pay for specific environmental control measures. Initial project profiles are developed and analyzed in critical sectors. Attention is focused on successive packages of investments needed to achieve both short- and long-term environmental goals.

- *Institutional feasibility.* This analysis looks at the institutional feasibility of carrying out alternative long-term environmental strategies. It concerns the policies, jurisdictional responsibilities, and organizational arrangements for environmental regulation and for the provision and production of environmental infrastructure and services. The influence of local institutional arrangements on the behavior of individuals, households, and enterprises also need to be understood. The assessment should identify the managerial improvements, organizational and jurisdictional changes, training requirements, and participatory mechanisms that would be needed to implement the strategy.

- *Policies and instruments.* This is an analysis of the policies and instruments needed to guide and motivate the behavior of those causing environmental problems and those responsible for controlling damages. The analysis focuses on matching policy instruments to the type of problem, the specific actor whose behavior needs changing, and the behavioral response called for by the strategy. Included are economic instruments, regulatory instruments, property rights, land management instruments, and information and education.

Box 4.3. The value of cost-benefit analysis for air pollution control

Deciding which environmental problems to address inevitably requires comparing the costs of damage with the costs of prevention. Where data are available, cost-benefit analysis can be used to compare policy options. An illustration comes from a recent study of air pollution in the southeastern Polish town of Tarnobrzeg. The economic benefits from reducing air pollution (lower mortality and morbidity rates, less material damage and reduced soiling) were compared with the costs of reducing emissions of total suspended particulates (TSP) and of sulfur dioxide.

The results were surprising—the benefits of reducing sulfur dioxide were exceeded by the costs in all cases. Yet, the benefits of reducing TSP by up to 70% exceeded the costs, with net benefits greatest for a reduction of about 40% (see figures below). Thus, measures to reduce TSP should be of higher priority in this region of Poland than those to control sulfur dioxide.

Total estimated benefits and cost of reducing exposure to air pollutants in Tarnobrzeg, Poland

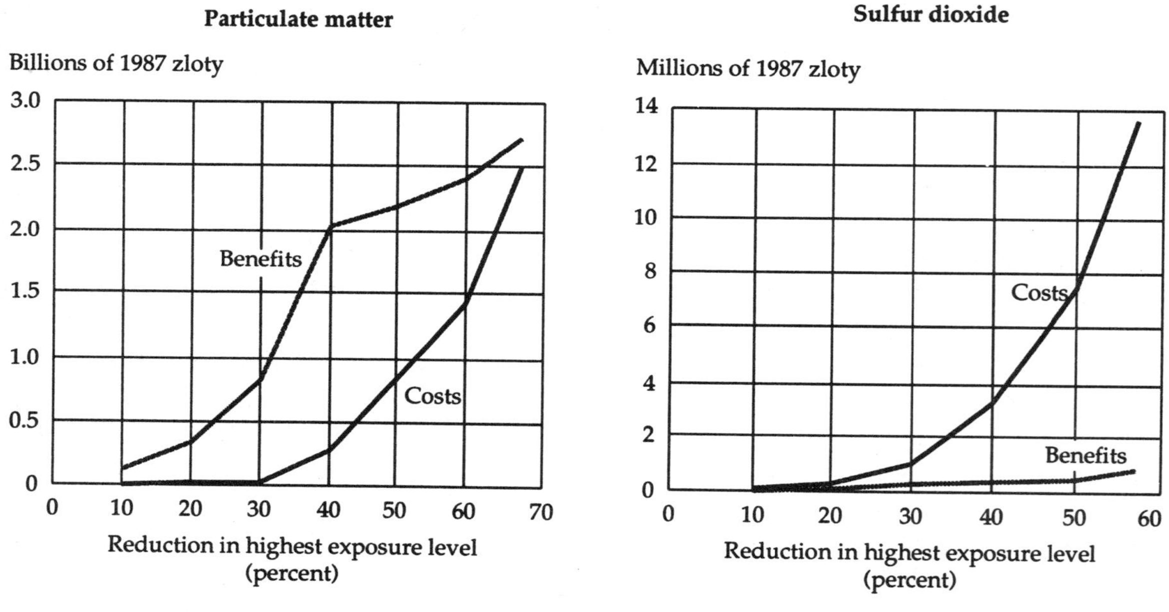

Source: World Bank *World Development Report 1992*

4.26 Given the extent of information needs and the complexity of the analyses to be carried out, the preparation of an issue-specific environmental management strategy for a large city will probably take six to nine months, perhaps more for a mega-city or metropolitan region, and up to several staff-years of effort. Factors that will affect both the time and resources required include the range of environmental problems to be dealt with, the agreed scope of the environmental management strategies and boundaries of the study area, and the degree to which information is available from previous planning studies.

Participatory approach for the environmental management strategy.

4.27 Broad-based acceptance of the strategy requires the participation of all stakeholders in the planning process and mechanisms for negotiation and conflict resolution. Although the preparation of the strategy will most often be led by public officials and technical specialists from the public and private sectors, a platform for public participation is essential. The exact type of organizational structure may differ among cities, but one proven approach is to create a steering committee and issues-specific working groups.

4.28 The steering committee (established during the consultation stage) would provide guidance on major project policies and issues, attain consent of those with legal responsibilities, and assure full coordination of effort and undertaking. The steering committee would include key decisionmakers from local authorities, and related national and regional sector agencies. To achieve consensus on the needed environmental improvements, the steering committee should also include leading politicians, community and business leaders, and representatives of important NGOs and the news media. If mobilizing international assistance is a goal, representatives from external support agencies could also be included.

4.29 Issue-specific working groups would be responsible for tackling specific priority problems. Each working group should include key stakeholders with direct access to the policy and decisionmakers in the public sector. The groups should also track the views and objectives of the private and non-governmental sectors through the steering committee and some form of a public advisory committee. Each working group would propose strategic options to deal with critical problems and analyze related human health, ecological, social, economic, and financial issues. Technical specialists supporting the working groups may be affiliated with central or local government institutes or agencies, consulting firms, academic or research programs, NGOs, or industry. External experts could also be drawn on as needed.

4.30 Many types of committee structures can work. For example, an arrangement in Leicester in the United Kingdom included eight thematic specialist working groups charged with developing Agenda 2020, a thirty year strategic plan. In the United States a three-committee structure was successfully used to tackle a multi-city, multi-state water pollution control problem in the Delaware estuary (see Case Study 4 in the Annex). In Jakarta a Steering Committee and a Technical Committee have been established to guide the preparation of an environmental management strategy being prepared with the support of the Metropolitan Environmental Improvement Program. In the case of Dar es Salaam, a consultation supported through the Sustainable Cities Programme led to the establishment of three inter-sectoral working groups and a technical support unit, a steering committee, and focal points in relevant organizations (see Case Study 1 in the Annex).

Coordinating an urban-wide environmental management strategy

4.31 An urban environmental management strategy needs to be developed in coordination with plans for urban development, for economic development, and for various types of urban infrastructure, energy development, and water resource management. Where these other plans do not exist or are incomplete or out of date, the environmental management strategy preparation is more difficult, but can provide the opportunity to initiate or update planning in other related areas.

4.32 A key output is an urban environmental management strategy report for use by local authorities, involved community groups, central and regional sector agencies, the business community, and others concerned with development and growth. The report should be viewed as a consensus document; the strategy it contains should guide future urban planning, economic sector planning, environmental services planning, natural resource and energy planning, and investments. The environmental management strategy report should summarize the following:

- A review of the environment/development issues, including a complete description of the environmental system from which each originates, the development concerns that are affected, and the stakeholders that should be involved.

- A summary of the possible responses to each issue, the pros and cons of each option for various stakeholders, and the assumptions and procedures that led to agreement on the selected options (including relationships and needed coordination with other issue-specific strategies, existing legislation, and ongoing programs).

- The set of agreed long-term environmental goals for the urban region and a set of interim objectives and targets to guide phased investments (this could include for each issue a ranking of environmental improvement measures, a preliminary outline of project profiles, and the identification of priority geographic areas and sectors for channeling investments).

- The associated policy reforms, economic and regulatory instruments, and institutional strengthening measures selected to support the implementation of the strategy.

4.33 The environmental management strategy should be a dynamic set of responses that take into account emerging environmental issues that may move up in priority as other problems are dealt with successfully. Another result, therefore, should be the institutionalization of the planning process. As the environmental management strategy guides planning, decisionmaking and sector interventions, the impacts of those decisions and interventions are periodically evaluated. This information is used in making incremental adjustments to the strategy as it evolves.

Actor-specific environmental action plans

4.34 To translate issue-oriented environmental management strategies into action plans specific actions must be defined for specific actors, complete with time schedules, geographic focus, and priorities. The goal is to provide the framework for coordination to ensure consistent actions across time, sectors and levels of government, and geographic areas. Emphasis is placed on implementing an investment program and on carrying out strategic policy responses.

4.35 Broad-based, informed participation and consensus continue to be essential for the successful formulation and implementation of the action plan. Effective participation can be facilitated through the same mechanisms as those used for agreeing on the environmental management strategy. The steering committee should review environmental action plans and resolve conflicts that may emerge. Working groups of stakeholders (supported by specialists) still have a role to play, although responsibility for the more technical work during action planning and project formulation is likely to shift from expected executing agencies to specialists. Actors that carry out projects may include municipal authorities and sector agencies, the private sector, NGOs, and CBOs.

4.36 To be effective, action plans should include a mix of mutually reinforcing instruments. The range of available instruments include regulatory mechanisms, economic incentives, technical assistance for capacity building, education and information campaigns, and strategic use of capital improvements.

4.37 Among the factors that will determine the success of strategic investments in urban environmental infrastructure and services, three stand out:

- *Standards and affordability.* Within a given city, and in a given sector, the environmental infrastructure or service to be provided can vary greatly. Technologies and service standards need not be uniform, but should be selected in accordance with users willingness and ability to pay. For example, some neighborhoods may demand full piped water supply and sewerage services, while other urban communities may only be willing to pay for yard taps and ventilated improved pit latrines. Form of payment may also be an important factor, whether in cash or in kind. Services should be based on effective demand to the greatest extent possible, allowing for service standards to vary across communities and for new, and socially acceptable, forms of service delivery.

- *Innovative institutional arrangements.* New institutional approaches may be called for in projects in view of local budget constraints, the recognized operational weaknesses of many local sector agencies, and the size of existing deficits of environmental services in most developing country cities. Any novel approach must respond to effective community demand for environmental services, and seek to incorporate cost-effective solutions. By wedding low-cost technologies with new institutional forms of participation, successful environmental interventions have been achieved in many cities, including community efforts in Olinda, Brazil and Orangi, Pakistan.

- *Cost-effective solutions.* Intervention options should be screened to determine their economic and financial soundness, and then should be ranked. A cost-benefit framework should be used whenever monetary costs and benefits can be estimated. In those instances where there is effective demand for environmental services, users' willingness to pay can be used to estimate benefits. Where only costs, but not benefits, are quantifiable, cost-effectiveness analysis can be used, based on the cost of the intervention and the incremental contribution that it would bring to the long-term environmental quality goals of the strategy (such as cost per ton of waste removed). This assumes a social willingness to pay for the environmental improvement, and is more likely to be valid where there is social agreement on the need to reduce major urban environmental spillovers.

Illustrations of the application of this type of cost effectiveness analysis to issue-specific action plans include water resource development in Beijing (Box 4.4) and air pollution control in Mexico City (Case Study 6 in the Annex).

Box 4.4. Cost effectiveness analysis of Beijing water supply options

When data are not available to undertake a full cost-benefit analysis, an assessment of the cost effectiveness of different environmental management options can be very useful. In the case of urban water supply, this involves comparing the costs of supplying (or conserving) various quantities of water, on a US$/m^3 basis, with the conventional cost of the next water development project.

This type of cost-effectiveness analysis was done for Beijing to evaluate the costs of conservation as an alternative to expanding water supply for the industrial and domestic sectors. The analysis shows that a variety of conservation measures in industries and households could release large quantities of water at a substantially lower unit cost than that of the next water development project.

Conserving water as an alternative to expanding supply in Beijing
(discounted cost in dollars per cubic meter)

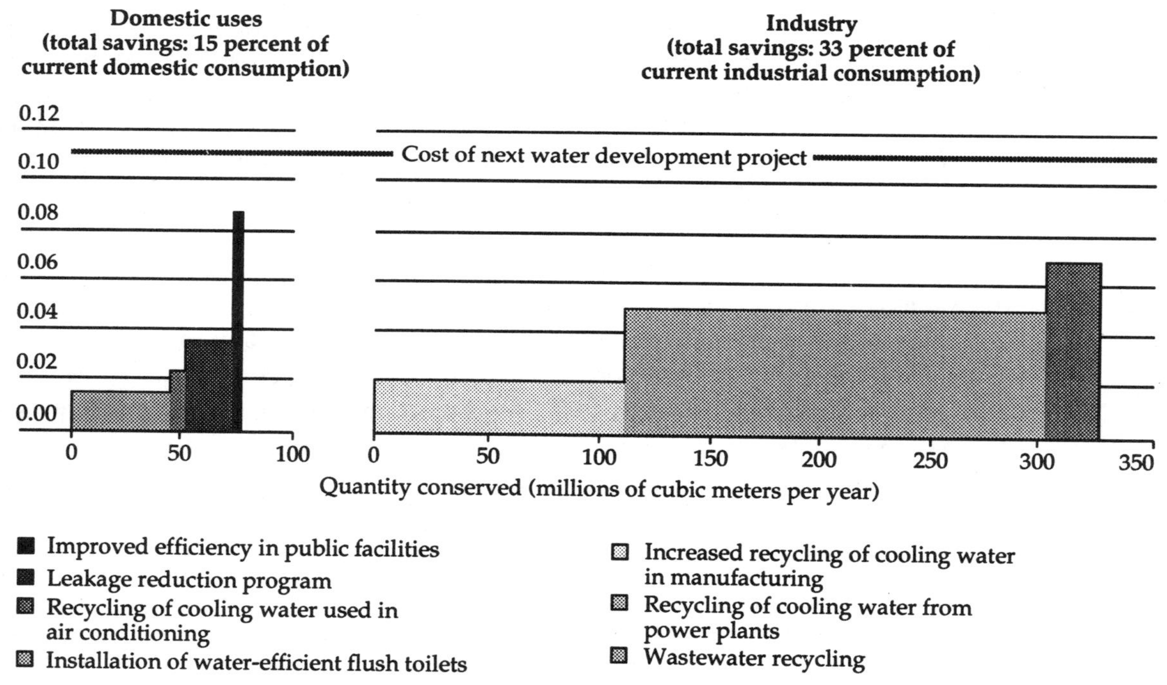

Source: World Bank *World Development Report 1992*

4.38 Interventions at the technical (projects) and operational level should be supported by corresponding actions at the institutional and managerial level and at the political level. The resulting action plans will focus on building institutional capacity in the following ways:

- effectively using of all available instruments for strategy implementation;
- devising mechanisms for coordinating actions across sectors and locations;
- cooperating effectively among the public, private, and community sectors; and
- making high quality information and technical support available.

4.39 Action plans are initially formulated for issue-specific environmental strategies, but sufficient coherency between plans must be achieved to ensure that they will form the building blocks of an urban-wide environmental action plan. The issue-oriented plans can be aggregated to identify all actions agreed upon by specific actors, thus providing clues about the institutional capacity of each actor and highlighting where more support may be needed.

4.40 When the issue-oriented action plans near completion (normally after nine to twelve months), the urban-wide environmental action plan can be formulated. This plan will comprise the set of specific actions that are needed to respond to priority environmental concerns in the city under study. Because these concerns are normally related to specific environmental media and spatial location, the plan focuses on the coordination and timing of various actions that are needed to handle critical problems. In addition to verifying that proposals are consistent with the goals and priorities of the environmental management strategy, the environmental action plan should also account for the costs and tradeoffs of competing sector actions. The plan should not be seen merely as the sum of individual actions taken by the different actors, but should also identify the specific cross-cutting actions required, specify who is responsible for implementing them, and pinpoint sources of funding. The key output here will be a set of action plans organized both by priority environmental issues and by responsible actors, and a consolidated environmental action plan. The issue-specific action plans include recommended projects and supporting actions. The environmental action plan report, in addition to summarizing the agreed strategies and actions, will describe the complementary cross-cutting actions and investment programs that have been agreed upon by participating stakeholders.

Follow-up and Consolidation

4.41 The final element in the urban environmental planning and management process extends over the time when agreed programs and projects are initiated, when policy reforms and institutional arrangements are solidified, and when monitoring and evaluation procedures are put in place. The focus here is on initiating programs and projects, building environmental planning and management capacities, and monitoring and evaluating progress. Depending on budget constraints and prevailing environmental conditions, a succession of staged investments spread over many years may be needed to reach the long-term environmental quality goals established in the environmental management strategy. Success will depend on sustaining investments, institutional strengthening programs, and policy reforms over time. Continuity is essential, since short- and medium-term gains can be easily lost by failing to follow through with needed actions in subsequent stages. Where continuity has been observed, as in the case of São Paulo, significant environmental improvements have also been evident (Box 4.5).

Box 4.5. External support of pollution control in São Paulo

A broad set of basic sanitation and pollution control investments have been undertaken by the government of São Paulo, primarily in the São Paulo Metropolitan area, over the past two decades. These efforts include World Bank-financed initiatives such the São Paulo Water Supply and Pollution Control Project signed in 1971, the Greater São Paulo Sewage Treatment Project of 1978, two sequential Industrial Pollution Control Loans signed in 1980 and 1987, and the Water Quality and Pollution Control Project for the Guarapiranga Reservoir in 1992. These investments, in addition to helping finance major pollution control facilities, also provided long-term support for important technical assistance measures such as institutional strengthening, training, and research.

The key pollution control agency in São Paulo has been the State Environmental Sanitation Technology Company (CETESB), which presently possesses the technical and administrative capacity, as well as the legislative mandate, to successfully oversee pollution control activities in São Paulo. Although CETESB was formally created in 1975, its foundations were laid with the establishment of the State Basic Sanitation Fund (FESB) in 1968 and a major UNDP/WHO technical cooperation project on pollution control in the early 1970s. Today, with a staff of over 2,100, including some 650 professionals, CETESB is widely recognized as the most experienced and technically capable environmental protection agency in Latin America. During the past decade, CETESB in turn has provided technical assistance to other state and local agencies responsible for the provision of environmental services in São Paulo, as well as in the rest of Brazil and elsewhere.

A recent evaluation of the two Bank-supported Industrial Pollution Control Loans concluded that the success of the program can be attributed to several key factors:

- existence of an adequate policy, legal, and regulatory framework at both the national and state levels;
- continual strengthening of an experienced institution (CETESB) that possessed both competent and highly motivated professional staff and the facilities and equipment required to properly monitor and control environmental degradation;
- effective use of a credit line for pollution control equipment and facilities and enforcement sanctions, including negative publicity, against polluters to induce them to use project or other resources to stem emissions or untreated effluents;
- strong political commitment to pollution control objectives by the state government after 1983, together with increasing public awareness of the health and other risks associated with growing pollution levels; and
- consistent Bank and other external support to CETESB, helping to reinforce its technical, administrative, and financial capabilities and improve its coordination with both private industry and other public sector agencies.

Source: World Bank (1992)

Initiating priority programs and projects

4.42 At this stage, proposed programs and initial project profiles developed in tandem with the action plan are advanced to the pre-feasibility stage to determine their ranking and their technical, economic, and institutional soundness. The size of the project(s) or investment package will depend on budget constraints and local implementation capacity. Detailed project designs and final appraisal of selected projects can then be completed. Preparation of "bankable" investment projects within an accepted project formulation framework is essential.

4.43 Financial and technical support for high priority projects can be mobilized through presentation and discussion of the environmental management strategy and the action plan with central ministries, national development banks, and external support agencies. A consultation is also advisable early on to brief potential national and international partners on the environmental management strategy and environmental action plan and to present priority investment packages and technical assistance requests.

4.44 Projects developed through this planning approach should more easily attract national and international support. The projects are specific and are formulated within a strategic framework that respects the cross-media, cross-sectoral, and cross-jurisdictional nature of environmental problems. They are products of a broad-based consultative process and have been endorsed by all those whose support is required for successful project implementation.

Creating a technical support unit

4.45 The only additional institutional facility to be created should be a centrally located technical support unit that serves as a focal point for organizing and technically supporting the concerted activities of the partners in the process. Public and private partners in the environmental planning and management process can relate to the technical support unit through designated focal points. Issue-specific working groups should still conduct most substantive work and decisionmaking. The steering committee should continue refining and monitoring the environmental planning process, and should be able to mobilize political support and negotiate cooperative arrangements among the key public and private sector actors in urban development.

Strengthening local capacity for environmental planning and management

4.46 The contribution of environmental planning and management efforts towards a healthier, more efficient, more productive, more equitable, and more sustainable city will depend largely on local absorptive capacity. The task of strengthening local capacities for urban environmental planning and management should be addressed in each stage of the planning process; the resulting institutional development recommendations will form part of the investment and implementation package. Capacity building activities can be initiated and supported by national and local governments as well as by external support agencies and international NGOs. Indeed, investments in institution-building have a high rate of return to donors. If such investments are effective, they result in new or more effective projects and services. This means quicker project implementation and, possibly, a steady stream of new projects once an institution is up and running.

4.47 ***Local capacity building.*** The capacities most critical to the success of environmental planning and management include adequate institutional arrangements and organizational structures, trained professional and support staff, adequate facilities and equipment, and resources to cover both capital and recurrent costs. These capacities need not be located in a single urban environmental management authority. Rather, the diverse cast of actors and institutions should work together to acquire these abilities, with each one building on existing structures and capacities to meet the requirements of environmental management.

External support for improving the urban environment

4.48 Because external support from donors for urban environmental management is limited and provides only a small portion of the assistance that is needed, such aid must be carefully leveraged. One way this can be done is by channeling donor assistance toward institutional strengthening measures. Donors can provide technical assistance and advice on management and administration and, in so doing, transfer valuable skills to urban administrators. But local initiative should always drive decisions about how to spend foreign aid or donor funds. Conversely, locally-developed urban environmental strategies and action plans should be used to guide the planning, implementation, and evaluation of interventions by external support agencies. By using these tools, external resources can be directed to the projects, services, or activities that have been identified by cities as priority investments for environmental improvement. The main types of assistance that cities can obtain from external support agencies include financial assistance for urban development and environmental projects, advice on policy and institutional reform, and support for research and policy analysis.

4.49 ***Support for urban development and environmental management projects***. Cities in developing countries can channel external resources to dedicated urban environmental management projects, to urban and infrastructure projects that include environmental components, and to institution-building efforts. In most cities, priority attention should be directed to urban waste management and pollution control facilities and services. In financing these improvements, cities and donors should ensure that investments extend waste collection services to the urban poor through increased efficiency and affordability. Investments also should include support for institutional strengthening (as needed) to ensure effective monitoring and enforcement of regulations, particularly pertaining to industrial and hazardous waste management. A key factor accounting for the success of São Paulo's industrial pollution control program, for example, has been the sustained external support to CETESB, São Paulo state's environmental company. In Indonesia, several donors have supported a government decision to establish a national environmental agency (BAPEDAL) that will have links to city-level environmental protection agencies.

4.50 The urban environmental management strategy will provide a framework for assessing the potential negative effects of infrastructure or other investments. The strategy will also include mechanisms for identifying alternative impact mitigation measures. Thus, it provides a way to meet the environmental impact assessment requirements of an externally funded project. Where there are potentially significant negative impacts associated with a project, the donor can either withhold funding for such a project or require the necessary impact mitigation measures and assist in their design.

4.51 ***Support for policy reform, institutional development, and resource mobilization.*** Despite the expansion of external financial assistance in recent years, the amount of aid financing will always remain small relative to investment needs. In most cities in the developing world, therefore, environmental problems will have to be solved through institutional and regulatory reforms and improved domestic resource mobilization, rather than solely or largely through foreign financial assistance. Even where external support agencies and international NGOs provide support, without major policy reforms, this assistance will not have a significant impact on the health and quality of life of the urban poor. The focus of donor support should thus be at the national level, where much of these major policy reforms originate. Extended effectively, such aid will then catalyze strategic environmental planning at the city level.

4.52 Cities in developing countries can obtain technical assistance and other aid from donors to get the environmental planning process off the ground. For example, the Urban Management

Box 4.6. Multilateral programs supporting urban environmental planning and management

• The **Urban Management Programme** (UMP), funded by the UNDP and a number of bilateral agencies and jointly executed by UNCHS (Habitat) and the World Bank, is a collaborative effort involving many national and local governments in the developing world. The program concentrates on strengthening the capacity of countries to address problems in five critical areas: urban land management; the provision and maintenance of urban infrastructure; municipal finance; the alleviation of urban poverty; and the protection of the urban environment. The environment component was initiated in 1990; its first phase involved research, background studies, and field work to define broad environmental management strategies for dealing with critical urban problems. In its second phase (1992-96), the component emphasizes support for (i) the preparation of urban environmental strategies in selected cities and countries; (ii) regional, national, and local capacity building for urban environmental management; (iii) high-priority research; and (iv) documentation of "best practice" approaches.

• The **Metropolitan Environmental Improvement Program** (MEIP), a UNDP-funded effort executed by the World Bank, initiated work programs in five Asian cities in 1990 (Beijing, Bombay, Colombo, Jakarta and Metro Manila). Its work focuses on the development and implementation of an environmental management strategy for each urban region. This includes activities to strengthen the capacity of pollution control and environmental protection agencies, especially in working with powerful economic planning and sectoral agencies at the local and national levels. MEIP also seeks to build a local environmental network, working largely with local organizations to do studies, demonstration projects and workshops on environmental problems and pollution abatement techniques.

• The **Sustainable Cities Programme** (SCP) was launched by UNCHS (Habitat) in 1990 to provide municipal authorities and their partners in the public, private and popular sectors with an improved environmental planning and management capacity. SCP demonstration projects result in broad-based environmental strategies, priority capital investment projects, and system-wide urban management capacities to mobilize all the public and private sector actors whose cooperation is required for successful implementation. As a global program, the SCP promotes the sharing of know-how between cities in different regions. As an inter-organizational effort, SCP mobilizes technical and financial resources from both bilateral and multilateral sources. The SCP is the global counterpart of the Asian MEIP and an operational arm of the UMP. City-level activities in various stages in the project cycle are under way in Accra, Concepcion, Dar es Salaam, Guayaquil, Ibadan, Ismailia, Jakarta, Katowice, Madras, and Tunis.

Programme, the Metropolitan Environmental Improvement Program, and the Sustainable Cities Programme are supporting environmental management strategies and environmental action planning activities in many cities (Box 4.6).

4.53 Given the weak and fragmented nature of both environmental and urban institutions in many developing countries, donors should provide countries with consistent advice on policy and institutional issues. At the municipal level, for example, the World Health Organization's Healthy Cities initiative seeks to commit urban managers to a concerted program of environmental health that is tailored to each urban area. The German Agency for Technical Cooperation (GTZ) is helping cities in Nepal and Thailand to implement urban environmental management guidelines. To ensure consistency, the international coordination of research and policy analysis, capacity building, and other reforms needs to be strengthened. Existing international networks that can be instrumental in promoting such coordination as and in facilitating city-to-city and community-to-community assistance for environmental management include the International Union of Local Authorities, the Habitat International Coalition, and other groups (Box 4.7).

4.54 ***Support for international research and policy analysis***. In light of the lack of global and city-specific data that are needed to formulate urban environmental strategies, donors should

Box 4.7. International networks supporting urban environmental planning and management

Numerous networks provide a range of assistance to municipal governments and city-based NGOs in the field of environmental management. Roughly, they can be divided between groups that link local governments with each other, and those that serve as a network for different types of city organizations. The following is a selected list of some of these networks.

Organizations of Local Governments

The **International Union of Local Authorities** (IULA) represents over fifty national associations of local government; it seeks to integrate environmental concerns into local planning and management, and increase the access of local and regional authorities to international organizations and meetings that deal with the environment. IULA, together with the UN Environment Program and the Center for Innovative Diplomacy, sponsor the **International Council for Local Environmental Initiatives** (ICLEI). ICLEI develops tools and management approaches for environmental protection, designs new approaches to difficult urban-centered environmental problems, and helps cities to develop local versions of the Earth Summit's Agenda 21. The **United Towns Organization (Cités Unies)** promotes twinning arrangements between cities for the transfer of environmental management skills and sponsors technical assistance for solving urban environmental problems. Other city-to-city organizations that are concerned with improving urban environmental quality include **Metropolis** and the **Summit of Major Cities of the World**.

Networks Linking City-Based Groups

Many international and regional organizations exist that link urban governmental and non-governmental entities in the area of environment. The **Habitat International Coalition** is an NGO network active in more than sixty countries that exchanges ideas and information on efforts to improve the quality of urban life. The **Mega-Cities Project** is a network coordinators in the world's most populous cities that documents, transfers and replicates innovations, including those of an environmental nature. Regionally, there are several networks functioning. In Asia, the UN sponsors **CITYNET** for Asian municipalities and **Asia-Pacific 2000** for NGOs. In the Middle East and North Africa, the **MEDCITIES** program (part of the Mediterranean Environmental Technical Assistance Program) fosters cooperation among Mediterranean cities. In Latin America, **CIUDAD** and **IIED-America Latina** gather and distribute information on urban environmental management. **ENDA-Tiers Monde** and **Environment Liaison Center International** (ELCI) link organizations and dissemination information in Africa and elsewhere.

Professional Organizations

Professional organizations provide additional opportunities for building urban environmental planning and management capacity in developing country cities. For example, the **International Solid Waste Managers Association** can be useful in promoting technology transfer, supporting the strengthening of national chapters and the development of local associations in developing country cities, promoting continuing education, sponsoring professional meetings, and publishing professional manuals on best practices in solid waste management. Similarly, the **International City Managers Association** promotes twinning arrangements, training, and technical assistance to developing country cities to improve waste management (sanitation, solid waste management) among other aspects of municipal government. Other key international associations for urban environmental management include the **International Water Supply Association** and the **International Association for Water Pollution Research and Control**.

support research on environmental conditions, impacts, problem causes, and the costs and benefits associated with alternative interventions. Areas where research is urgently needed include: health impact estimation, economic valuation, environmental management techniques, and social assessment. More broadly, the collection of environmental data is needed in all sub-sectors. To ensure some degree of consistency in approaches and comparability among countries, all research, data collection, or policy analysis by the support agency should be well-coordinated with other relevant donor activities and with existing regional or sub-regional urban research networks. Agencies supporting research should also ensure that the results of the various research and policy analysis activities are disseminated to urban authorities, academic researchers, international urban networks, and other donors. Two new financing mechanisms administered by the UNDP to support such research and analysis (as well as demonstration activities) are the Local Initiative Facility for Urban Environment (LIFE) and the Global Environmental Facility Small Grants Program.

Monitoring and evaluating progress

4.55 Performance indicators must be monitored and evaluated over the long-term and mechanisms must be set up for feedback from the implementation experience. In this way the steering committee can systematically analyze, assess, and adjust the environmental planning and management process. The results should also feed into associated data collection and information analysis. Such monitoring and evaluation is particularly important in rapidly developing economies where policies, institutional arrangements, and social and economic conditions are changing rapidly.

4.56 Regular monitoring and evaluation are needed so that emerging urban environmental issues are recognized and addressed before they become costly emergencies, and so that existing environmental management strategies and action plans can be routinely assessed and adjusted. To ensure effective implementation of strategies and plans, monitoring and evaluation should focus on:

- monitoring of performance indicators for environmental infrastructure and services;
- monitoring of implementation of prevention and mitigation measures; and
- long-term ambient monitoring to guarantee compliance with environmental standards.

4.57 Capacity building activities should be routinely monitored and periodically adjusted to reflect changing needs. In evaluating existing programs, local authorities should assess whether all critical environmental issues have been identified. This can avoid costly crises. Checks should also be made on whether management strategies and action plans have been agreed upon for all identified issues, and whether all relevant policy instruments available to the partners in the environmental planning and management process are being effectively used. Based on an assessment of environmental management in Japanese cities, the factors that contributed to successful implementation include: rapid economic growth, political and economic equity, integration of industrial promotion and environmental management, decentralization of important government functions to competent local governments, and an educated, articulate population (Ex Corporation 1992).

ANNEX A: CASE STUDIES

Case Study 1: Environmental Consultation in Dar es Salaam

Background

A1.1 In early 1991, the Government of Tanzania asked to participate in the UNCHS (Habitat) Sustainable Cities Programme (SCP). Preparatory activities of the SCP included: (a) a city environmental profile that organized available information on environmental, developmental and institutional conditions; (b) a formal project agreement with associated funding; and (c) a carefully prepared and structured consultation that involved all project partners so that they could set priorities, clarify roles, establish procedures, and develop a sense of common purpose and ownership. The preparatory activities also involved *ad hoc* operational support to meet high- priority needs identified through the SCP. The support was mobilized in collaboration with the UMP and included updating base maps from satellite imagery, organizing an emergency solid waste clean-up campaign, and providing legal and technical advice on privatizing waste collection.

The consultation

A1.2 A five-day city consultation from August 23-September 1,1992 was a collective achievement. It was based on information in the environmental profile and structured according to a generalized SCP format (the "consultation shell") that was adapted to the specific circumstances and priorities in the city. In attendance on each of the five days were more than 100 participants from the public, private and community sectors, including ten cabinet ministers with their principal secretaries, members of parliament, city councilors, business people, community activists, journalists, and local experts, all of whom contributed to plenary and group discussions. The consultations received widespread media coverage.

A1.3 Concrete results from the consultation included: (a) clarification of priority urban environmental issues; (b) the establishment of inter-sectoral working groups, a multi-disciplinary technical support unit, a steering committee, and focal points in relevant organizations, all for further SCP activities; (c) the commitment of technical staff and an office for the SCP; and (d) lessons for conducting future city consultations. Results from the consultation will contribute to the preparation of a city-wide environmental management strategy and the development of an urban environmental action plan. Some of these results are summarized in the Dar es Salaam Declaration (see Box A1).

Follow-up

A1.4 To ensure continued momentum, technical continuity, and long-term follow-up, the consultation marked the beginning of the environmental Strategy and Action Planning Phase. This phase is expected to result in the following outputs: (a) information and technical analyses to clarify the environmental issues being adressed; (b) issue-specific environmental management strategies agreed upon among all those whose support and cooperation is required for their effective implementation; (c) a rolling overall urban environmental management strategy (EMS), to replace the outdated master plan; (d) an urban environmental action plan (EAP) involving concerted interventions at operational, institutional, and political levels; (e) preparation of technical coopera-

Box A1.1. Dar es Salaam declaration

The Consultation on Environmental Issues for Managing the Sustainable Growth and Development of Dar es Salaam (26th August to 1st September 1992)

Recognizing that deteriorating environmental conditions in and around the rapidly expanding and densifying city of Dar es Salaam present major obstacles to achieving sustained and equitable socio-economic growth and development;

Accepting that efficient and effective environmental management for the city of Dar es Salaam is based on the following principles: (1) that natural resources are the basis for sustainable development, and that environmental hazards threaten development achievement; (2) that the environment forms an integral concern in development management; (3) that environmental issues cross development sectors, geographic space and time; (4) that environmental management must reconcile competing interests; (5) that environmental management entails coordination, implementation and technical support; (6) that environmental management must adapt to and involve existing mechanisms and institutions; (7) that environmental management must evolve incrementally over time; (8) that environmental management requires active involvement of those whose interests are affected and whose support is required; and (9) that environmental management requires deliberate and continuous public awareness and political support;

Having deliberated on a number of key environmental issues and discussed the papers and case studies presented on specific problems of solid waste management and servicing urban land in the city; we the participants of this consultation have gained a better understanding of each others' roles and responsibilities in city management, not only across sectors, between levels of government, but perhaps most importantly with respect to the contributions that both the private sector and community based organizations can make to more effective urban environmental management;

Recommend that: 1. The Dar es Salaam City Council cannot continue to be the provider of subsidized services due to the limited central government financial subsidy and poor revenue collection;

2. A new partnership approach to city management is therefore required, which broadens the range of actors involved to include the private sector and community based organizations;

3. Working groups be established to address the key environment issues identified, which must have representation from (a) affected sectors and levels of government, (b) affected geographic locations, and (c) the private sector, and community groups and interested individuals in order to overcome the poor coordination arrangements which are seen to represent the principal constraint on effective city management so that they may together chart out concerted actions in sufficient operational detail to ensure the sustainable growth and development of our city recognizing its unique history, cultural and social background, natural environment and economy;

4. A clear focal point be established through the Sustainable Dar es Salaam Project where the partners can feel comfortable to meet and deliberate upon issues, and discuss alternative courses of action if such working groups are to be effective;

5. Such a focal point be directly integrated with existing city management practices, but with sufficient independence to encourage a sense of ownership and participation by the city's partners in the public, private and community sectors so that they may readily contribute their ideas, information, skills and financial resources;

6. The Sustainable Dar es Salaam Project focuses its resources on building up a core of dedicated interdisciplinary professionals from both levels of government, the private sector and community organizations who will together support the planning, coordination and management of the city's future growth and development in a sustained manner requiring in turn (a) the necessary supporting facilities, equipment and services, (b) secondment by relevant institutions through the Ministry responsible for Local Government of additional professionals to participate in the project technical support team and working groups, with commensurate incentives; (c) to nominate desk officers to act as project focal points within the key government agencies;

7. Broader policy issues be reserved for central government decision, especially in the field of urban land, allowing the city authorities to be the active implementors of such policies in planning the city's expansion, detailed subdivision design, survey and land allocations;

8. Integrated cross sub-sectoral working groups immediately be established for the following two priority issues and topic areas: (a) solid waste management, (b) servicing urban land;

9. Additional working groups be established through a consultative process similar to this consultation to address environmental issues such as the following: (a) air quality management and urban transport, (b) surface water management and liquid waste, (c) management of coastal area resources, (d) management of recreational resources and tourism attraction, and (e) management of urban agricultural potential;

10. The above be addressed through the Chairman of the concluding session, the Hon. Attorney General, to the Hon. Minister of State for Local Government, requesting that this declaration be forwarded to the Right Hon. Prime Minister for his information, and to all Ministers for their cooperation, information and continuing support. Copy to the Resident Representative of the United Nations Development Programme and the Executive Director of the United Nations Centre for Human Settlements (Habitat) for their information and continuing support; and copied to all participants.

(102 signatures)

tion and capital investment projects; and (f) support to develop environmental planning and management capacity. During a three to six month follow-up and consolidation phase, priority projects will be initiated, policy reforms and institutional improvements will be solidified, and procedures for monitoring and evaluating the effectiveness of the EPM process will be put in place. In the end, after a two-year SCP demonstration, an urban environmental planning and management process should have been established.

A1.5 Since the consultation, the municipal government and its partners in the public and private sectors have begun to develop and apply new approaches to human settlements planning and management, with an explicit emphasis on sustainable urban development. This effort originally developed around the priority issues identified during the consultation: solid waste; servicing urban land; air quality; surface water; coastal area resources; recreation and tourism resources; and urban agricultural potential. New urban management techniques were introduced and operationalized through the process of addressing these issues. The approach has suceeded in widening the basis of participation in development decisionmaking and in mobilizing a wealth of local resources through new partnerships among the public, private, and community sectors.

A1.6 By late 1993, the cross-sectoral and inter-institutional planning and management arrangements established around the initial priority issues have become a framework through which additional urban environmental issues are now being addressed in a more coherent and effective manner. Additional technical cooperation support (approximately $15 million) for building local urban environmental planning and management capacity has been arranged with nine bilateral and eight multicultural external support agencies. Stimulated by these achievements, the Government of Tanzania has intiated a national programme for sustainable urban management to replicate the lessons learned in Dar es Salaam in Tanzania's eight intermediate cities. The Dar es Salaam municipal government and the Government of Tanzania are being supported in these efforts by UNDP's country programme (about $1 millon for Dar es Salaam and $2 millon for the national programme), organized through the Sustainable Cities Programme.

Source: Jochen Eigen "City Consultation on Environmental Issues," draft article for *HABITAT NEWS*, November 1992.

Case Study 2: Rapid Urban Environmental Assessment in the Côte d'Ivoire

A2.1 Management of urban environmental problems was selected as one of three key themes for the Côte d'Ivoire urban sector review at the end of 1991, along with poverty alleviation and land management. An inter-ministerial Environment Working Group was established to assess the status of environment in the urban sector, and develop recommendations for improving environmental management in the capital and secondary cities. This work is being done in coordination with two other working groups, as well as with the team that will develop Côte d'Ivoire's national environmental action plan.

A2.2 The environmental working group identified ten areas of environmental concern for the country's cities:

- surface water pollution and supply
- solid waste management
- sanitation
- urban transport
- open and green space
- preservation of cultural heritage
- energy use and supply
- management of urban markets
- industrial pollution
- land use.

The working group also developed a tentative list of options for addressing these environmental problems. However, they realized that issues and options needed to be further developed, ranked, and placed in the overall context of the urban sector review.

Scope and objectives

A2.3 A three-step process (described below in more detail) is being used to more fully and quickly assess the environmental issues and options facing cities in Côte d'Ivoire. The process is being used by the UNDP/UNCHS (Habitat)/World Bank UMP, and by the UNDP/World Bank Metropolitan Environmental Improvement Program (MEIP) and the UNCHS SCP. This approach is designed to gather relevant information on the state of the urban environment in Côte d'Ivoire; to understand the interactions and institutions that affect urban environmental quality; to rank problems; and to outline a sub-sectoral strategy.

A2.4 The environmental working group's efforts will be based on an analysis of Abidjan and seven secondary cities. The secondary cities were chosen according to geographic location, population, growth rate, and range of environmental problems (see table A2.1). The cities and criteria are summarized in the table below. Four of these cities (Bouaké, San Pedro, Ferké, and Bouna) are being used as case studies by the other two working groups. Eight other cities may be added with the help of U.S. Peace Corps volunteers working on an urban environmental management project.

Table A2.1. Selection of secondary cities

City	Location	Population[1]	Rate[2]	Environmental problems
Bouaké	C	332,000	5.2	squatter settlements, grey water, industrial emissions
Gagnoa	W	95,000	5.5	drainage, marketplace, water quality, roads
San Pedro	SW	82,000	7.5	sanitation, drainage, land subsidence, housing
Abengorou	E	65,000	5.0	water management, marketplace, forest degradation
Grand Bassam	SE	45,000	3.6	coastal erosian, historical heritage, eutrophication
Ferké	N	37,000	2.6	industrial pollution, desertification, water
Bouna	NE	15,000	4.8	drought, housing

1. 1990 estimate
2. average annual growth rate, 1973–88

Process

A2.5 ***Information gathering: environmental indicators questionnaire.*** A questionnaire on urban environmental indicators has been developed by an international technical working group and field-tested by the UMP. It covers baseline social and economic statistics; housing and health conditions; the natural environment; land use; urban transport; urban energy use; air pollution; noise pollution; water resources, water supply and sewerage/sanitation; and solid and hazardous wastes. The questionnaire can be completed relatively quickly by local consultants (using secondary sources) in conjunction with the environmental working group. For Abidjan, a full set of environmental data have been collected using the indicators questionnaire; for each of the secondary cities, a reduced set of data were assembled (tables on background statistics, natural environment and land use, air pollution and energy use, and water resources and urban waste).

A2.6 ***Understanding the setting: environmental profile.*** Information from the questionnaire, along with relevant studies and other documentation, will be used to prepare a four-section urban environmental profile. First, the background section provides a historical, geophysical and socio-economic perspective on the urban environment. Second, the status section summarizes existing information on the quality of environmental media (air, water, land, and cultural property), and briefly analyzes key natural hazards (both geogenic and human-induced). Third, the development-environment section describes how development-oriented activities and services in the public, private and informal sectors influence environmental quality, and how environmental factors constrain or promote development. Fourth, the setting section identifies the key public and private actors engaged in environmental management, the existing management functions (instruments of intervention, and mechanisms for coordination and decisionmaking), and the initiatives that are being undertaken to improve environmental management. A full profile has been prepared for Abidjan; an executive summary, covering these four sections, was written for each of the secondary cities.

A2.7 ***Establishing priorities: the consultation process.*** The profile is useful for delineating the nature and extent of urban environmental problems. But this precedes ranking, which requires public discussion and technical judgment. Several consultations (consisting of individual interviews or small roundtable meetings) will be organized with municipal politicians, local government planners, environmentalists, regional/national officials, community groups, universities, NGOs, and private industry. These interviews will explore different perspectives on environmental priorities, constraints, and opportunities. Participants will then convene at a "town meeting" on the urban environment to try and reach a consensus on priorities. The working group would then use these results, along with technical inputs, to establish a ranking that balances public perception and scientific knowledge. With baseline information, an understanding of key interactions, and a sense of priorities, the working group should be able to sketch a general environmental management strategy for tackling key urban environmental problems (see Case Study 3).

Case Study 3: Environmental Management Strategy for Metropolitan Manila

Background: rationale and process

A3.1 The Government of the Philippines is extremely concerned about the increasing rate of environmental degradation in major cities throughout the country, especially in the Metro Manila Region (MMR). The MMR consists of seventeen separate local government units that make up a total land area of 636 square kilometers and a 1990 populated estimated at 9.8 million. To begin to contain urban pollution, the Government started in September 1991 to prepare an environmental management strategy for the MMR, linking it to the preparation of an industrial efficiency and pollution control program. The objective of the study was to prepare a broad pollution control strategy for the MMR to provide a framework for public and private sector actions.

A3.2 The strategy study was initiated by its two principal client agencies with the Government—the Department of Environment and Natural Resources and the Department of Trade and Industry. It was funded by the UNDP/World Bank Metropolitan Environmental Improvement Program and was undertaken by a consortium of two consulting firms (one Filipino and one foreign), with advice from a local technical working group and oversight from a local joint steering committee. Work proceeded in three phases leading to: (a) an inception report (November 1991) listing preliminary findings and problems encountered; (b) a problem identification report (January 1992) that served as a technical working paper for discussion with (among others) a technical working group and a joint steering committee; and (c) an interim report (May 1992) presenting draft findings and recommendations. The final report on the strategy was issued in August 1992; its recommendations are now being reviewed by the Government and donors.

Findings: environmental conditions and trends

A3.3 Generally, the strategy found that deteriorating urban environmental conditions have accompanied rapid population growth and expansion of the MMR, with development occurring in a largely random, unplanned way because of a lack of appropriate land management policies. The conditions of different environmental media are as follows:

- *Water.* All surface water is contaminated with human, solid, and industrial wastes to varying degrees, with the major river systems classified as biologically dead. Lake water quality is still acceptable for aquatic life and industrial uses, but may be threatened by hazardous wastes. The aquifer level has been lowered by over-extraction and degraded by saltwater intrusion. The coastal waters of Manila Bay are considerably polluted, with high dissolved oxygen levels and fecal/total coliform counts. For water pollution, the key sources of biological oxygen demand are domestic wastewater (40 percent), industrial wastewater (38 percent) and solid waste (22 percent). Motor vehicles are the largest source of most air pollutants followed by industries and utilities, though the bulk of particulate matter consists of resuspended soil.

- *Land.* Inadequate solid waste management is a major cause of land and water pollution, with 1,000 tons of solid waste going uncollected every day. Settlement patterns are uncontrolled as is industrial estate development.

- *Air.* The MMR has some of the worst urban air pollution in the world. High concentrations of particulates and lead pose serious threats to human health.

A3.4 If no action is taken, the quality of life will continue to deteriorate and public health will be adversely affected. With increasing sedimentation in Laguna Lake and the continued dumping of heavy metals in Manila Bay, fisheries could be wiped out and tourism would decline. With ground water levels receding dramatically and with saline intrusion near the shoreline, the natural supply of ground water could be exhausted, requiring substitution with high-cost alternative water supply sources.

Elements of a priority-led strategy

A3.5 Using previous studies and new information on rates of environmental degradation, the following ranking (in descending order of importance) of urban environmental priorities was developed:

(a) Solid waste
(b) Flooding
(c) Toxic/hazardous pollution
(d) Slum/squatter settlements
(e) Water pollution
(f) Air pollution
(g) Degradation of natural resources
(h) Noise pollution.

Several sub-strategies were then designed to deal with these priorities. Solid waste, flooding, and water pollution problems are tackled in a water quality management strategy; hazardous wastes are included in a toxic and hazardous substances strategy; issues involving slum/squatter settlements and degradation of natural resources are included in a land use management strategy; and air pollution is the focus of the air quality management strategy. Issues of human and financial resources that cut across all the strategies are tied together in strategies for institutional strengthening and financing.

A3.6 The water quality management strategy consists of: (a) an accelerated septic tank desludging program; (b) expansion of a combined sewer system through the Second Manila Sewerage Project; (c) development of two sanitary landfills and five transfer stations; and (d) waste minimization and wastewater treatment for the top 100 polluting firms. The hazardous waste strategy calls for provision of temporary on-site storage facilities and development of a centralized treatment and disposal facility. The land use management strategy seeks to incorporate environmental issues into land use plans, along with development of a plan for greening of the MMR. The air quality management strategy begins with a program to phase lead out of gasoline and to reduce the sulphur content of fuel; it also includes a program for vehicular and industrial emissions monitoring. The strategy for institutional strengthening includes reform that clearly designate the Metro Manila Authority as the key power for metropolitan land use, measures to improve the monitoring and enforcement capacities of the Department of Environment and Natural Resources and the Laguna Lake Development Authority. The strategy also includes support for training, for political consensus-building, and an information/public education campaign. The financing strategy proposes both an environmental management project for external funding, and the establishment of an Environmental Management Fund supported by discharge fees paid by industries.

Investments and returns

A3.7 The overall strategy would cost $186 million. The major components, in descending order of cost, are: wastewater treatment, waste minimization for existing industries, waste minimization for new firms, hazardous waste management, pollution abatement appraisals, air pollution control equipment, institutional strengthening, air emission monitoring, and a business center. Additional programs would be supported with local resources through the Environmental Management Fund. If the project is not undertaken, the estimated economic and social costs of continued urban environmental degradation would amount to P1.23 billion (about $53 million) in the year 2000. By that time, the annual net environmental benefits associated with the project (in the year 2000) will be an estimated P732 million (about $32 million).

The knowledge gap

A3.8 The study also identified priority areas where more information and analysis are needed. One of the top priorities is the improvement of surface water quality models; they need to be expanded, modified, or replaced. Ground water contamination needs to be assessed; once data are available, a study to evaluate the institutional structure, resources and regulations for ground water management is needed to recommend appropriate changes. And environmentally acceptable land use plans and zoning ordinances should be developed, along with training for relevant staff.

Case Study 4: Water Quality Management in the Delaware Estuary

A4.1 This case study presents the concepts of a participatory planning approach for developing an issue-specific environmental management strategy affecting several cities—in this case, for regional water quality management in the Delaware Estuary. The example is based on an innovative planning effort carried out for the Delaware Estuary in the mid-1960s for the purpose of setting water quality goals.

Background

A4.2 With its headwaters originating in the state of New York, the Delaware Estuary is an 86-mile stretch of tidal river that begins at Trenton, New Jersey, and flows by the Philadelphia, Pennsylvania and Camden, New Jersey area as well as several intermediate cities such as Wilmington, Delaware. Federal, state, inter-state, and local level governments are involved in its management, primarily through the Delaware River Basin Commission.

A4.3 When the study was conducted in 1960, water quality in the estuary was already poor, and recreational uses were severely curtailed. Commercial fishing was estimated at only about 8 percent of possible capacity, with frequent fish kills caused by low oxygen conditions in the estuary. Thus, the economic losses associated with poor water quality were significant, and the aesthetic value of the estuary was also depressed. Most cities in the area did not disinfect their waster discharges, giving rise to high summer bacteria levels and concern about health risks.

Planning approach

A4.4 The interest-based planning approach that was adopted was designed to take into account the following considerations:

- the difficulty of determining cause and effect relationships between waste discharges and water quality, which requires a rigorous technical analysis;
- the (usually) poorly articulated nature of water quality goals;
- the ease with which a realistic and reasonable range of objectives can be obtained from a regional group of water-using interests;
- how, by having a range of objectives at hand, one can evaluate the consequences of the goals under discussion and continually provide information to water-using interests, regulatory, and technical interests;
- the importance (as the analysis continues) of asking water-using interests for recommendations on a single water use objective; and
- the usefulness of presenting all information and recommendations (at the conclusion of the analysis) to appropriate decision-making bodies for action and implementation.

A4.5 The basic premise of the interest-based planning approach is that if an efficient communications network can be established between water-using groups, a set of objectives will emerge that can form a meaningful point of departure for decision-making. Applying this to the Delaware case, a tripartite advisory committee structure was adopted. It consisted of a Policy

Advisory Committee, a Technical Advisory Committee, and a Water Users Advisory Committee (WUAC).

A4.6 The Policy Advisory Committee included representatives from state, interstate, and federal agencies having the legal power to abate pollution. The committee attained consensus from the various state and other government agencies on pollution abatement policies and plans. Committee members worked to ensure the inclusion of established water quality plans in the overall water pollution control strategy.

A4.7 The Technical Advisory Committee included experts familiar with the technical aspects of water pollution control and representatives from agencies participating in water pollution control work. The technical committee kept these agencies apprised of the technical status of the study, provided technical assistance to the federal government in the planning and coordination of the study, helped organize sub-projects, and advised the Policy Advisory Committee and WUAC on technical matters.

A4.8 WUAC members were drawn from all of the water-using interests in the study area. The WUAC advised study participants on the water use and water quality needs and desires of the people in the study area. The committee also served a public relations function. The parent committee was formed by combining four subcommittees: a general public subcommittee, an industrial subcommittee, a local government subcommittee, and an environmental interest subcommittee. Although the general public and environmental interests were not as well organized and did not have the same motivations as the major water withdrawal and waste discharging interests, they had equal status in the parent committee.

A4.9 The work of the WUAC was initiated by formal meetings held with each of the four subcommittees. At opening meetings the philosophy, objectives, and methodology of comprehensive planning were explained and the role of the WUAC was reviewed in detail. Each group then selected a chairman to sit on the WUAC to represent their subcommittee.

A4.10 The WUAC's effort to determine the water use and quality desires of each of the subcommittees was divided into two phases. In Phase I the committee elicited from each group what they saw as the water use and quality needs and desires they wanted to achieve. Phase II was intended to provide more information on the specific location of present and desired future water uses and specific ranges and/or values of individual water quality parameters associated with each water use.

Water use and water quality objective sets

A4.11 Based on the work of all the committees, five water use/quality objective sets were prepared. Each objective set delineated various reaches of the estuary where different water uses would be made suitable from a water quality standpoint. The uses considered included municipal and industrial water supply, sport and commercial fishing, and water-contact recreation. In summary, the five sets were:

- *Objective Set 1 (OS-1):* This set would provide the greatest increase in water use and quality. Water contact recreation would be made suitable in the upper and lower ends of the estuary (about forty-five miles would be suitable for swimming). Sport and commercial fishing opportunities were placed at relatively high levels. The passage of anadromous fish was included as a definite part of the water quality management program.

- *Objective Set 2 (OS-2):* The area of water contact recreation was reduced somewhat from the first set (about one-third of the estuary would be suitable for swimming). A reduction in the dissolved oxygen goal was considered to result in a concomitant reduction in sport and commercial fishing. Anadromous fish passage remained as a specific aim of the management program.

- *Objective Set 3 (OS-3):* This was similar to OS-2. The management for anadromous fish passage was not included as a definite goal of this set, although substantial increases in passage would result from treatment requirements imposed to control oxygen levels during the summer. A general decrease in sport and commercial fishing potential was imposed through a lowering of the oxygen requirements.

- *Objective Set 4 (OS-4):* This set represented a slight increase over 1964 levels of water contact recreation and fishing in the lower reaches of the estuary.

- *Objective Set 5 (OS-5):* This set represented maintenance of 1964 conditions, that is, a prevention of further water quality deterioration.

A4.12 For each of the objective sets, twelve water quality parameters were considered including dissolved oxygen, coliform bacteria, turbidity, and temperature, among others. This linkage between water use and water quality then formed the quantitative basis for estimating the effects of various waste reduction programs.

Waste reduction programs—costs and benefits

A4.13 The various objective sets were analyzed in terms of the type of point waste source reduction program that would be necessary to achieve the water quality goal. Specific attention was directed to dissolved oxygen. As there were several waste sources and reduction options, optimum waste control models were formulated under assumptions of overall cost minimization of treatment, uniform treatment throughout the estuary, or zoned treatment. A water quality model was developed and used as the basic cause/effect mechanism to relate waste load discharge to resulting water quality.

A4.14. The results of the application of the management models provided information on the effectiveness of options and the sensitivity of the cost of solutions to such physical parameters as dispersion and re-aeration, as well as providing a basis for estimating the costs of various administrative control policies. Total costs and marginal costs were estimated for each of the objective sets and under the three control programs. Benefit estimates were also made for each of the objective sets. Figure A4.1 plots the total costs and quantifiable benefits on an ordinal abscissa scale, including the costs and benefits of the program finally chosen by the Delaware River Basin Commission. The maintenance of 1964 water quality conditions in the face of increased raw waste

Figure A4.1. Costs and benefits associated with water quality objective sets in the Delaware Estuary

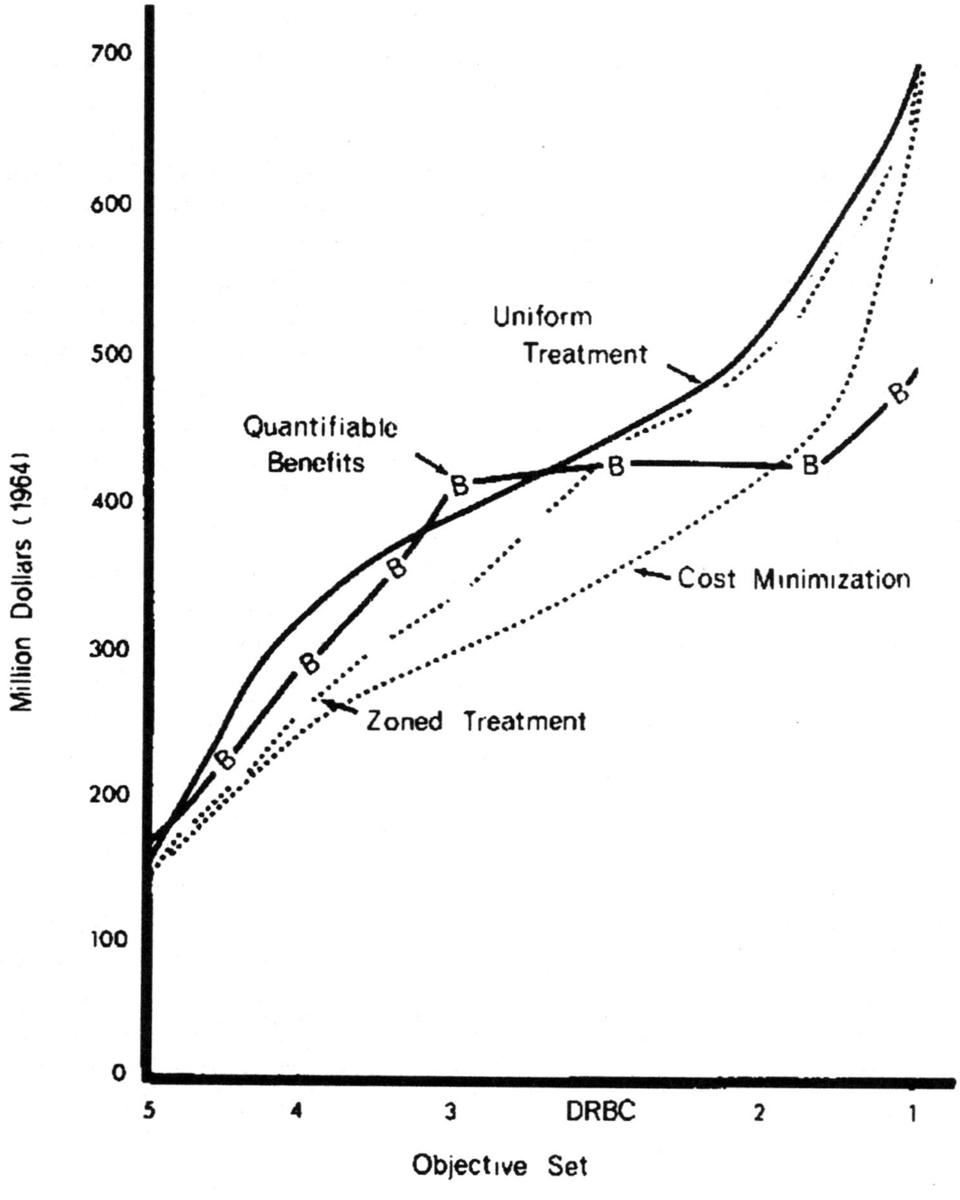

loading through 1975-80 would require an expenditure of about $140 million (1964 dollars). The difference in cost between the uniform treatment option and the cost minimization option for the lower and middle objectives was substantial and averaged about $90 million. After deliberation among the committees and following an analysis of marginal costs and benefits, it was decided that a set of objectives similar to OS-3 appeared most desirable from a quantitative analysis.

The decisionmaking process

A4.15 Next, the five objectives were formalized into complete sets with accompanying control programs and presented to the WUAC for specific recommendations to the Delaware River Basin Commission. Ultimately, the WUAC in March 1966 agreed to recommend Objective Set 3 as its goal for the Delaware Estuary. This was a result of three subcommittees preferring OS-3 while the fourth group (the environmental interests group) preferred OS-2.

A4.16 A series of public hearings, discussions, and dialogue commenced after presentation of the preliminary results and water quality objectives by the Delaware River Basin Commission at a conference in 1966. It became clear that the policy and administrative sectors favored a higher objective that OS-3, the set chosen by a majority of the WUAC. However, at the annual Delaware River Basin Commission governing board meeting in March 1967 OS 2 was selected as the goal for the Delaware Estuary. This choice (with some modification), led to widespread frustration among the industrial and local government groups of the WUAC. These groups felt that, after the considerable effort and expense of participating in the decision making process, the recommendations and concerns of the group were completely discounted in the final hours. In reality, much of the input of these water use groups formed the framework for the range of objectives used in the decision making process. The concept of desirable water uses, leading to specification of water quality levels and then to waste load discharge requirements, formed an important part of later public hearings. At the very least, because of the participation of the entire estuarine community, a setting was established within which each interested group had played a role. The implementation of the Delaware River Basin Commission quality program, albeit not the program desired by the industrial and municipal community, incorporated several unique features in water quality management that benefitted all concerned.

Epilogue

A4.17 Having set the water quality objectives, the Basin Commission established a series of zones and arrived at a uniform percent removal for each zone. A reserve of about 10 percent was set aside as a hedge against uncertainty about future discharges. All dischargers are required to install a minimum level of nominal secondary treatment (about 90 percent). For each zone, the percent removal is applied to the raw load of each discharger to determine his waste load allocation. This bypasses the question of when a load becomes significant to the water quality of the estuary. Primary attention from an enforcement point of view is paid to the large municipal and industrial discharges. The burden of sampling for compliance is on the discharger.

A4.18 This assignment of waste discharge allocations to achieve and maintain a water quality objective represented a significant departure from standard U.S. regulatory practice. Each waste discharger has full knowledge of what is expected as opposed to vaguely stated treatment requirements or broad scale effluent concentration criteria. As time progresses, the percentage of waste removal required along the Delaware will have to increase substantially. For some zones, advanced treatment achieving greater than 95 percent removal will be required for organic wastes and nutrients by the year 2000 to maintain the allocations and water quality objectives.

Source: Thomann, Robert J. "Comprehensive Water Quality Management—the Delaware Estuary." In *Systems Analysis and Water Quality Management*. New York: McGraw-Hill, 1972.

Case Study 5: Catchment Protection in the São Paulo Metropolitan Region

A5.1 The formulation of a watershed protection plan for the São Paulo Metropolitan Area (SPMA) is a good example of an issue-specific urban environmental management strategy. It started with a rapid assessment to gauge explicit needs and inventory problems. Consensus among the various actors was achieved through consultations, culminating in an issue-specific strategy. The plan set the stage for coordinated short- and medium-term sectoral action plans and investments.

The approach taken

A5.2 In the São Paulo approach, once priorities were set and consensus was achieved, a series of sectoral action plans and investments were designed. The lesson from São Paulo is that such a consensus-based strategy should comprise the following:

- the agreed long term environmental goals for the urban region;
- a set of interim environmental goals and objectives to guide phased investments;
- the ranking of pollution control and other measures to improve environmental quality;
- the identification of priority sectors for channeling investments, including project profiles; and
- the recommended policy reforms, instruments, and institutional development needed to implement the environmental management strategy.

Background

A5.3 The greater SPMA is the most urbanized, industrialized and affluent region in Brazil. It consists of thirty-eight cities, with São Paulo City (SPC) being the largest (current population is 11.4 million. The SPMA is already one of the largest and fastest-growing urban regions in the world, having a population approaching 20 million and registering a growth rate averaging nearly 5% annually since 1960. Population is projected to be 24 million inhabitants by the year 2000 by which time the SPMA will be the second largest urban agglomeration in the world after Mexico City. The SPMA covers 8,051 square kilometers, including a metropolitan core of 900 kilometers. The region, with 12 percent of Brazil's population, produces nearly a third of the country's industrial output and accounts for 18 percent of gross domestic product (Leitmann 1991).

A5.4 Further expansion of the SPMA is bounded on the north and south by major water basins that are the sources for most of the urban water supply. Urban growth, however, has been in a southerly direction, resulting in water quality deterioration in the Guarapiranga Reservoir, which serves 30 percent of the urban population. Guarapiranga Basin's population, now 492,000, is growing at 3 percent a year. Sixty-five percent of the inhabitants are located around the reservoir itself, and most of the others live in three adjacent municipalities in the upper catchment area. Approximately 24,500 families live in *favelas* without urban services. The wastes from these communities are discharged directly into the reservoir along. Upstream municipalities are sewered and have solid waste collection services, but are prohibited by federal law from discharging wastes within the catchment area. Compliance is difficult because neighboring municipalities outside of the catchment area have passed local bylaws prohibiting the importation of wastes from other jurisdictions. Federal land use restrictions in the catchment area are so severe that landowners do not develop their properties, leaving them vulnerable to invasion by squatters to form new favelas. Agricultural and industrial activities in the catchment area are also problematic and better controls are needed.

A5.5 Water quality of the Guarapiranga reservoir has been seriously compromised as a result of these problems. For example, over the past decade nutrient discharges (nitrogen and phosphorus) have increased, resulting in the proliferation of algae in the reservoir. This in turn has led to a doubling of treatment costs, obstruction of treatment filters, and taste and odor problems. The water quality problems are well-publicized and widely recognized by the urban population.

The São Paulo consultation

A5.6 To mobilize support around concern for water quality problems and to back up that concern with hard data an initial consultation was combined with the preparation (by local consultants) of an environmental profile of São Paulo. At the initial consultation it was unanimously agreed that Guarapiranga required urgent action. The profile, which included interviews with city, community, and business leaders, was followed up by a town forum on the urban environment. Participants included over 120 city and state government officials, members of NGOs and community groups, academics and researchers, leaders of professional organizations, and members of the press. Among the priority issues flagged in relation to water quality in the Guarapiranga watershed were substandard housing, lack of environmental infrastructure and services for the poor, and settlement on risk prone areas.

The issue-specific environmental management strategy

A5.7 In 1991 a working group was set up with a goal of devising a solution to the problem. The group, led by the State Water and Sanitation Company (SABESP), includes among others the State Environmental Agency (CETESB), the Metropolitan Planning Authority (EMPLASA), and the four municipalities involved in the Guarapiranga Basin. An environmental strategy was formulated for the basin and an action plan prepared and presented to the World Bank for financing. The Bank participated in and supported all phases of this process. The action plan will cost an estimated $229 million, and includes the following elements:

Immediate urban areas:

- land use rationalization and control, including lot readjustment and urban rehabilitation;
- resettling favelas and low-income households located in risk-prone and geologically unstable areas;
- access roads and urban infrastructure;
- integrated provision of water supply, sewerage, drainage, slope stabilization, and solid waste collection in presently unserved areas; and
- elimination of direct waste discharges and runoff into the reservoir by interception and pumping to an adjacent basin (the Tiete River) where future wastewater treatment will be provided.

Upper catchment area:

- facilities for upstream municipal wastewater treatment and solid waste disposal;
- land use management;

- development of recreational parks and ecological areas as buffer zones around the reservoir shoreline;
- control of erosion and nutrients from agricultural runoff; and
- restrictions on mining and industrial activities.

A5.8 Responsibility for these actions is spread across several state and municipal authorities, each of which has made a commitment to participate and finance their share of the agreed actions. Key instruments to be applied so as to meet the objectives of the project and ensure the desired benefits include:

- a surcharge for Guarapiranga water use by São Paulo consumers;
- adequate pricing policies for resources and services in the catchment area;
- increased property taxes and application of betterment levies to all local beneficiaries;
- development of recreational parks to protect critical and damaged areas and prevent future occupation by favelas, managed by private concessions;
- integrated pollution control actions across state agencies and municipalities; and
- training programs for watershed management, including technical and financial support to NGOs and the four municipalities.

Conclusion

A5.9 One of the main medium-term institutional reforms would be to create a Guarapiranga water basin management agency. The action plan calls for studies and the organization of the public and governmental consultation process needed to develop and establish the basin agency. The agency will have primary environmental planning, monitoring and enforcement functions in the basin. An indirect benefit of the project will be to develop a modern scheme of river basin management, which, in the medium term, would be expanded on a larger scale to the other major basins in the region that serve the SPMA as well as nationally.

Source: Bartone, Carl and Emilio Rodriguez. "Watershed Protection in the São Paulo Metropolitan Region: A Case Study of an Issue-Specific Urban Environmental Management Strategy." *Infrastructure Notes*, Urban No. UE-9. World Bank Transportation, Water and Urban Development Department. Washington, D.C., April 1993.

Case Study 6: Mexico City's Plan for Tackling Air Pollution

A6.1 The Mexico City Metropolitan Area (MCMA) is one of the most populous urban areas in the world and has the worst air pollution of any city in the Western Hemisphere. MCMA's population has grown from less than 1 million at the beginning of the century to 15 million in 1990, and is projected to reach 20 million by 2010. The critical air pollutants in the MCMA are lead, ozone, carbon monoxide (CO), and fine particulate matter. Peak ozone concentrations, the highest of any city in the world, are more than four times the accepted national and international norms. Peak period total suspended particulate levels are among the highest recorded in any city in the world, and far exceed the Mexican health advisory level. CO far outweighs all other pollutants combined in terms of total mass of emissions, but levels are less of a threat to health, because CO is much less hazardous than other pollutants. The major source of atmospheric lead in the MCMA is combustion of lead anti-knock compounds in gasoline.

The action planning process

A6.2 Given the urgency of the air pollution problem in the MCMA, the World Bank and the Mexican Government undertook a sector study, drafted in 1991, to provide the technical and analytical underpinnings for a proposed Bank project on transport air quality management in the MCMA. The study was prepared by Bank staff and consultants, and draws heavily on work conducted by the Mexican Government in connection with its "Programa Integral Contra la Contaminacion Atmosférica de la Zona Metropolitana de la Ciudad de México" and work carried out by a joint team of Mexican and international consultants for the Departmento del Distrito Federal. Development of the action plan involved: (a) estimating health and economic impacts of air pollutants, based on sources, total emissions, toxicity, and exposure patterns; (b) evaluating the Government's current air quality strategy for the MCMA; (c) assessing transport policy options, based on their cost-effectiveness; and (d) designing a strategy that includes institutional improvements.

A6.3 ***Estimating health and economic impacts.*** Quantitative data on the health and other economic impacts of air pollution in the MCMA were only beginning to be developed at the time of the study. Preliminary analyses suggested that excess levels of fine particulate matter may be responsible for 12,500 extra deaths and 11.2 million lost workdays a year, while ozone may account for 9.6 million lost workdays per year, both due to respiratory illnesses. Around 140,000 children were estimated to suffer a reduction in IQ and agility, requiring remedial education due to excessive lead exposure. About 46,000 adults suffer from hypertension for the same reason, and about 330 of these adults die of heart attacks. Additional heart attacks and other circulatory problems can also be expected due to CO exposure, but data on actual exposures were insufficient to provide a quantitative estimate. Total economic damages due to the health effects of air pollution in the MCMA were conservatively estimated at $1.5 billion per year.

A6.4 ***Evaluating the government's air quality strategy.*** Since 1988, air pollution control programs have been greatly expanded in the MCMA. An Emergency Air Pollution Control Program was put into effect in 1989 and consists of: adoption of tighter vehicle emissions standards, a rotating one day a week driving ban, a mandatory reduction in gasoline lead content, a vehicle inspection and maintenance program, upgrading of bus engines, and expansion of the air quality monitoring network. In September 1990, the Mexican authorities announced an expanded effort, the Integrated

Program Against Air Pollution in the Mexico City Metropolitan Area. It includes forty-two specific measures affecting the following areas: (a) oil industry and fuels; (b) transport sector; (c) private industry and services; (d) thermoelectric plants; (e) reforestation and sanitation; and (f) research, education and communication. (A revised version was issued in April 1991.) The initiative was effectively coordinated and implemented; however, air quality in the MCMA has not improved. It is unclear whether this was due to shortfalls in the program and/or to unfavorable weather conditions. The program should result in a substantial reduction in overall pollutant emissions. It could be improved with better information, planning, coverage, clearer targets and priorities, more strident economic incentives, better enforcement and monitoring, and direct linkage to transport sector investments and policies.

A6.5 One lesson from the tepid results of the program was that regulations alone can backfire—as did banning each car on a specific day of the week in Mexico City. Rather than stay off the streets, many drivers bought a second car. The added cars pushed up the costs of administration. And, with more cars in the city, many of them older, the regulation ended up increasing pollution, not reducing it. This perverse outcome shows that understanding the effects of regulations and all other incentives is crucial for sound policy-making.

A6.6 ***Assessing transport policy options.*** The study began by considering the following policy goals: (a) to limit vehicle emissions per vehicle-kilometer (such as by use of catalytic converters); (b) to make better use of existing road space (such as by increasing vehicle occupancies); (c) to influence the relative use of public and private transport; (d) to reduce vehicle-kilometers by limiting the use of private vehicles or shortening travel distances; (e) to reduce total passenger-kilometers; and (f) to minimize motorized travel. A decision was made to focus on measures to limit emissions per vehicle-kilometer travelled because such reductions require straightforward technical (rather than social or behavioral) changes. The following options to reduce transport emissions were identified: fuel pricing; emissions standards for new vehicles; reformulation of fuels; gasoline vapor recovery; segregated transit; improved bus service; road infrastructure upgrading; freight redistribution; and a range of demand management measures.

A6.7 Ideally, the ranking and evaluation of these alternative air quality measures would be done by comparing the costs and benefits of each, and selecting those offering the greatest net social benefit. However, quantitative data linking pollutant emission reductions with human health and economic impacts were not yet available. Thus, the options were assessed in terms of their cost-effectiveness (cost per kilogram of emissions eliminated). As figure 6.1A reveals, some technical measures—such as retrofitting cars to use liquefied petroleum gas and making service stations recover vapors—yield net savings. Others ride up the marginal cost curve. Emission standards, applied to different groups of vehicles, are at the lower end. Mandatory inspection and maintenance programs for vehicles are in the middle—rising as the standards are strengthened. Improvements in the fuel mix are at the higher end.

A6.8 Beyond $1,200-$1,500 a ton, however, emission reductions get very expensive. At that point, all that remains is demand management—to keep people out of their cars. Picking the best from a menu of instruments without a gasoline tax could reduce emissions by 50 percent in 1995—at a cost of $560 million. Adding a gasoline tax to the menu would allow the elimination of some costly initiatives, and the same emission reductions would cost $460 million—and generate $300 million in public revenues as an added benefit.

Figure A6.1. Cost of controlling air pollution from transport in Mexico City

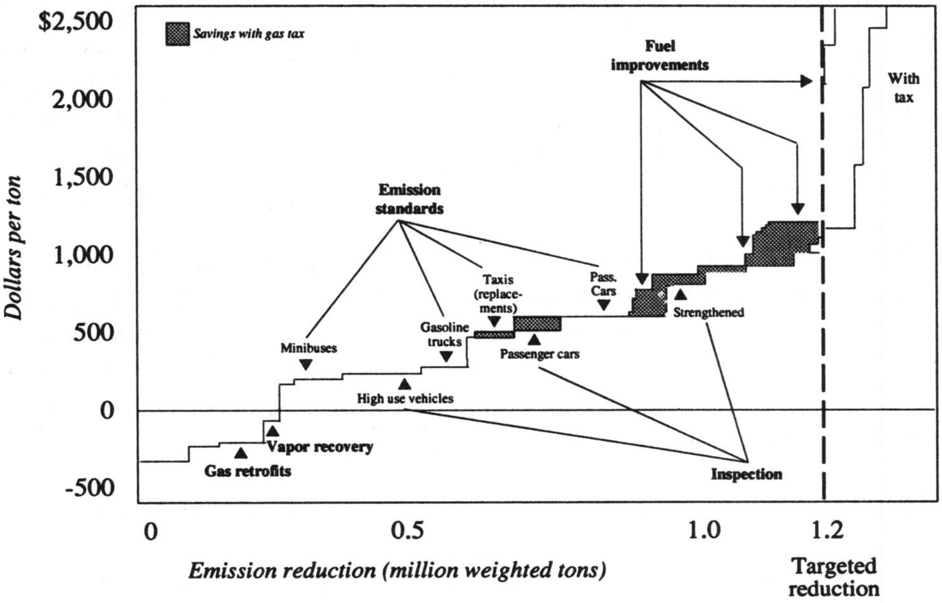

Source: Gunnar Eskeland, World Bank, 1992.

A6.9 ***Designing and financing a strategy.*** The stated purpose of the proposed Transport and Air Quality Strategy for the MCMA is not to replace individual sub-sector strategies, but to refine, extend, and unify them. The fully-developed strategy will consist of a balanced package of policies that will involve vehicles, fuels, transportation policy, and research and education. In addition, it will create and strengthen institutions needed to implement the strategy, monitor its effectiveness, and carry out further air quality planning. Provision will be made for the mobilization of resources to carry out the strategy. Thirty-four strategy components have been identified, based on least-cost analysis.

A6.10 Many of these components are now being integrated into a World Bank-supported project, Transport Air Quality Management Project for the Mexico City Metropolitan Area, which is under appraisal. Tentatively, it will consist of the following:

- *A vehicle component* including (i) support to develop, promulgate and enforce emission standards for new vehicles, (ii) progressive improvements to in-use emission standards, and inspection and maintenance programs for high-use vehicles and private cars, (iii) lines of credit to finance the replacement of old, high-use vehicles (taxis, trucks and minibuses) with new, emission-controlled vehicles, conversions of high-use vehicles to liquefied petroleum gas (LPG) or compressed natural gas (CNG), and re-engineering and retrofit of emissions controls, and (iv) improvements to the vehicle registration system;

- *A fuel component* to help fund the installation of vapor recovery systems at service stations, and an alternative fuel pilot program to speed the conversion of high-use vehicles to LPG or CNG;

- *A transport policy and management component* to help prepare an integrated transport and air quality management strategy for the MCMA, including travel demand management, urban freight management, public transport, and transport investment planning;

- *A scientific base component* to help fund consultant services to strengthen air quality planning in the MCMA through an integrated research plan and equipment for extending the air quality monitoring system, short-term pollution forecasting and the mobile-source emissions laboratory; and

- *An institutional strengthening component* to support agencies involved in air pollution control as well as independent annual environmental audits.

A6.11 ***Tackling the institutional issues***. The planning and implementation of a program to control air pollution in the MCMA is greatly complicated because the problem has national, regional and local characteristics. As a result, a multitude of federal, state and municipal agencies are involved in policy making; effective implementation and coordination are difficult to achieve. Since the MCMA includes both the Federal District and seventeen municipalities in the state of Mexico, and air pollution originates in and affects the entire Valley of Mexico, the problem is regional in scope, requiring common action across jurisdictional boundaries. However, such areas as vehicle inspection, traffic management and public transport, for example, are the responsibility of individual agencies in the federal, state and some municipal governments.

A6.12 To address these cross-sectoral and cross-jurisdictional issues, the Government created the Commission for the Prevention and Control of Environmental Pollution in the Metropolitan Area of the Valley of Mexico (CMPCCA) in 1992. The CMPCCA has responsibility for the prevention and control of all aspects of environmental degradation in the region (with an initial focus on air pollution) by defining and coordinating the policies, programs and projects of public agencies at all levels of government. The World Bank project will provide CMPCCA with technical assistance to improve its scientific base and strengthen its technical support team. In the longer term, the CMPCCA will need to develop a sustainable source of funding, possibly via revenues generated through the

ANNEX B: ENVIRONMENTAL STRATEGIES WORKING DOCUMENTS

A number of interim papers and reports were prepared in part to contribute to the development of this strategy paper. Some of these have been published by the Urban Management Programme, some are available in draft form from the UMP, and others have been published elsewhere.

UMP Papers

Energy-Environment Linkages in the Urban Sector (UMP Paper #2).

Alternative Approaches to Pollution Control and Waste Management: Regulatory and Economic Instruments (UMP Paper #3).

A Review of Environmental Health Impacts in Developing Country Cities (UMP Paper #6).

Satellite Remote Sensing and GIS Analysis for Urban Applications (UMP Paper #9).

Environmental Innovation and Management in Curitiba, Brazil (UMP Working Paper #1).

Land Use Considerations in Urban Environmental Management (UMP Paper #12).

Rapid Urban Environmental Assessment: Lessons from Cities in the Developing World. Volume 1—Methodology and Preliminary Findings (UMP Paper #14).

Rapid Urban Environmental Assessment: Lessons from Cities in the Developing World. Volume 2—Tools and Outputs (UMP Paper #15).

Economic Valuation of Environmental Improvements (UMP Paper #19). *Public-Private Partnerships for Municipal Solid Waste Management. Volume 1—The Private Sector* (UMP Paper #13).

Unpublished Draft UMP Documents

"Urban Environmental Data Questionnaires and Profiles of Accra, Jakarta, Katowice, Sao Paulo, the Singrauli region of India, Tianjin, and Tunis." 1991.

"Priorities for Urban Waste Management and Pollution Control in Developing Countries."1992.

"The Sustainable Cities Consultation Shell." 1993.

Conference Papers

Bartone, C.R. "Addressing Environmental Issues in Urban Areas." Second International Seminar on the Environmental Problems of Urban Centers - ECO URBS'93, São Paulo, December 12-17, 1993.

Bartone, C.R. and L. Benavides. "Local Management of Hazardous Wastes from Small-scale and Cottage Industries." Fifth Pacific Basin Conference on Hazardous Wastes, Honolulu, November 8-12, 1993.

Bartone, C.R., J. Bernstein, and J. Leitmann. "Responding to the Environmental Health Challenge of Urbanization in Developing Countries." 20th Annual NCIH International Health Conference on Health and the Environment: Meeting the Challenge for Human Development, Arlington, VA, June 20-23, 1993.

Bartone, C.R. "Economics and Finance of Urban Sewerage Services in Developing Countries." USETI Course on Industrial and Municipal Wastewater Treatment Facilities, Washington, DC, May 10-21, 1993.

Bartone, C.R., J. Bernstein, and J. Leitmann. "Managing the Environmental Challenge of Mega-Urban Regions." AIT International Conference on Managing the Mega-Urban Regions of ASEAN Countries: Policy Challenges and Responses, AIT, Bangkok, November 30-December 3, 1992.

Bartone, C.R. "Environmental Management Strategies for Cities." Salzburg Seminar on Protecting the Urban Environment, April 6-16, 1992.

Leitmann, Josef. 1992. "Rapid Urban Environmental Assessment: A First Step Towards Environmental Management in Cities of the Developing World." Twelfth Annual Meeting of International Association for Impact Assessment, Washington, DC, August 19-22, 1992.

Leitmann, Josef, C. Bartone and J. Bernstein. 1992. "Environmental Strategies for Developing Country Cities." International Workshop on Planning for Sustainable Urban Development, University of Wales, Cardiff, July 13-17, 1992

Journal Articles

Amuzu, A.T. and J. Leitmann. 1994. "Environmental Profile of Accra, Ghana." *CITIES* X (1).

Bartone, C.R. and J. Bernstein. 1993. "Improving Municipal Solid Waste Management in Third World Countries." *Resources, Conservation and Recycling* 8(1):43-54.

Bartone, C.R. 1991. "Institutional and Management Approaches to Solid Waste Disposal in Large Metropolitan Areas." *Waste Management & Research* 9(6):525-536.

Bartone, C.R., L. Leite, T. Triche, and R. Schertenleib. 1991. "Private Sector Participation in Municipal Solid Waste Service: Experiences in Latin America." *Waste Management & Research* 9(6):495-509.

Bartone, Carl. 1991. "Environmental Challenges in Third World Cities." *Journal of the American Planning Association* 57(4):411-415.

Hadiwinoto, Suhadi and J. Leitmann. 1994. "Environmental Profile of Jakarta, Indonesia." *CITIES* X (3).

Leitmann, Josef. 1994. "Environmental Profile of Katowice, Poland." *CITIES* X (3).

Leitmann, Josef. 1993. "Rapid Urban Environmental Assessment: Toward Environmental Management in Cities of the Developing World." *Impact Assessment* 11 (3): 225-260.

Leitmann, Josef, C. Bartone & J. Bernstein. 1992. "Environmental management and urban development: issues and options for Third World cities." *Environment and Urbanization* 4 (2): 131-140.

Leitmann, Josef. 1992. "Rapid urban Environmental Assessment." *FORUM Valutazione* 4(1): 47-69.

Oliveira, Celso and J. Leitmann. 1994. "Environmental Profile of São Paulo, Brazil." *CITIES* X (2).

Other Publications

Bartone, Carl and E. Rodriguez. 1993. "Watershed Protection in the Sao Paulo Metropolitan Region: A Case Study of an Issue-Specific Urban Environmental Management Strategy." *Infrastructure Notes* No. UE-9.

Leitmann, Josef. 1992. "Satellite Remote Sensing and GIS Analysis for Urban Applications." *Infrastructure Notes* No. OU-6.

Leitmann, Josef. 1992. "Tianjin: Towards an Improved Urban Environment." *The Urban Age* 1(1).

Leitmann, Josef. 1992. "A Tool for Rapid Urban Environmental Assessment: The Urban Environmental Profile." *Infrastructure Notes* No. UE-8.

Leitmann, Josef. 1992. "A Tool for Rapid Urban Environmental Assessment: The Urban Environmental Indicators Questionnaire." *Infrastructure Notes* No. UE-5.

McNeil, Mary. 1991. "The Consultative Process in Setting Urban Environmental Priorities." *Infrastructure Notes* No. UE-4.

REFERENCES

Ahmad, Yusuf J. 1986. "Saving the Pearl." *The Siren*, Number 32, November.

Amuzu, A.T. and Josef Leitmann. 1991. "Environmental Profile of Accra." Draft Urban Management Programme Case Study, August.

Bartone, Carl. 1990. "Sustainable Responses to Growing Urban Environmental Crises." Paper presented at International Union of Local Authorities (IULA) World Development Forum, Brussels, 4-6 April.

Bartone, Carl, Janis Bernstein, and Frederick Wright. 1990. "Investments in Solid Waste Management: Opportunities for Environmental Improvement." World Bank Policy Research Working Paper Series No. 405, Washington, D.C.

Bartone, Carl, Luiz Leite, Thelma Triche, and Roland Schertenleib. 1991. "Private Sector Participation in Municipal Solid Waste Services: Experiences in Latin America." *Waste Management and Research*, Volume 9.

Bernstein, Janis. 1991. *Alternative Approaches to Pollution Control and Waste Management: Regulatory and Economic Instruments.* Urban Management Programme Discussion Paper Series No. 3.

———. 1993. *Land Use Considerations in Urban Environmental Management.* Urban Management Programme Discussion Paper No. 12. Washington, D.C.: World Bank.

Bertaud, Alain. 1992. Written communication, April.

Bhatnagar, B. and A.C. Williams. 1992. *Participatory Development and the World Bank.* World Bank Discussion Paper No. 183. Washington, D.C.: World Bank.

Borkiewicz, Jerzy, Ewa Mieczkowska, Alicja Aleksandrowicz, and Josef Leitmann. 1991. *Environmental Profile of Katowice.* Urban Management Programme, Draft Case Study.

Bradley, David, Carolyn Stephens, Trudy Harpham and Sandy Cairncross. 1992. *A Review of Environmental Health Impacts in Developing Country Cities.* Urban Management Programme Discussion Paper No. 6. Washington, D.C.: World Bank.

Briscoe, John. 1990. "Brazil: The New Challenge of Adult Health." World Bank Report No. 7807-BR. Washington, D.C.

Bower, Blair T., Charles N. Ehler, and Allen V. Kneese. 1977. "Incentives for Managing the Environment." In *Environmental Science & Technology*, Volume 2, No. 3.

Castaneda, Fernanco Casas. 1989. "The Risks of Environmental Degradation in Bogota, Colombia." *Environment and Urbanization*, Volume 1, No. 1, April.

Chemical Manufacturers Association. 1991. "Responsible Care: A Public Commitment."

Chia Lin Sien. 1979. "Coastal Changes, Marine Pollution, and the Fishing Industry in Singapore." In *Economics of Aquaculture, Sea-Fishing and Coastal Resource Use in Asia.* Eds. Aida R. Librero and William L. Collier. Manila: Philippine Council for Agriculture and Resources Research.

Clarke, Giles, Suhadi Hadiwinoto and Josef Leitmann. 1991. "Environmental Profile of Jakarta." Urban Management Programme Draft Case Study.

Cointreau, Sandra. 1987. "Solid Waste Management Study for the Greater Banjul Area, The Gambia." Final Report sponsored by The Gambia Government, Ministry of Economic Planning and Industrial Development.

Cointreau-Levine, Sandra. 1994. *Private Sector Participation in Municipal Solid Waste Services in Developing Countries.* Urban Management Programme Discussion Paper No. 13. Washington, D.C.: World Bank.

Convery, F.J and K.A. Tutu. 1990. "Evaluating the Costs of Environmental Degradation in Ghana." Report Prepared for the Ghana Environmental Protection Council. Accra, Ghana.

Dagh-Watson, SPA. 1987. "Present Waste Management and Resource Recovery, Both Formal and Informal, Within the Metropolitan Area of Mexico City." Draft Final Report, UNDP/ World Bank Resource Recovery Project, World Bank.

Diaz, Virginia Jiminez. 1992. "Landslides in the Squatter Settlements of Caracas; Towards a Better Understanding of Causative Factors." Environment and Urbanization. Vol. 4, No. 2. October.

Di Pace, Maria, Sergio Federovisky, Jorge E. Hardoy, Jorge E. Morello, and Alfredo Stein. 1992. "Latin America." In *Sustainable Cities: Urbanization and the Environment in International Perspective.* Boulder: Westview Press.

Dowall, David E. and Giles Clarke. 1991. *A Framework for Reforming Urban Land Policies in Developing Countries.* Urban Management Programme Paper No. 7. Washington, D.C.: World Bank.

Economic Commission for Latin America and the Caribbean. *1991. Sustainable Development: Changing Production Patterns, Social Equity and the Environment.* Santiago: Economic Commission for Latin America and the Caribbean.

Eigen, Jochen. 1992. "City Consultation on Environmental Issues." *HABITAT NEWS*, September.

Environmental Law Review Committee. 1991. "Environmental Legislation: The Increasing Costs of Regulatory Compliance to the City of Columbus." Report prepared for the Mayor and City Council of the City of Columbus, May.

Environmental Quality International. 1988. "Extension of Technical and Advisory Services to the Zabbaleen Gameya-Phase II. Report presented to the Ford Foundation, December.

Eskeland, Gunnar S. 1992. "The Objective: Reduce Pollution at Low Cost." *Outreach #2*. World Bank Country Economics Department, June.

Ex Corporation. 1992. "Japan's Experience in Urban Environmental Management: Interim Report." Prepared for UNDP/World Bank Metropolitan Environmental Improvement Program. Tokyo, December.

Farvacque, Catherine and Patrick McAuslan. 1992. *Reforming Urban Land Policies and Institutions in Developing Countries*. Urban Management Programme Discussion Paper Series No. 5. Washington, D.C.: World Bank.

Ford Foundation. 1993. "Urban Research in the Developing World." Reports presented to the Final Meetings, Cairo, February 14-18.

Foster, J. David. 1992. "The Role of the City in Environmental Management." Prepared for USAID Office of Housing and Urban Programs Workshop. Bangkok, Thailand, Septemeber.

German Agency for Technical Cooperation (GTZ). 1993. "Manual for Urban Environmental Management." Eschborn: Urban Environmental Guidelines Project.

________. 1992. "Urban Environmental Guidelines for Nepal." Kathmandu: Urban Development Through Local Efforts Project.

Golden, Tim. 1993. "After the Deluge, Tijuana's Future is Dimmer." *New York Times*. February 1.

Grenon, Michel and Michel Batisse. 1989. *Futures for the Mediterranean Basin—The Blue Plan*. New York: Oxford University Press.

Hahn, Ekhart and U. Simonis. 1991. "Ecological Urban Restructuring." *Ekistics*. Volumes 348-349.

Hamza, Ahmed. 1990. "An Appraisal of Environmental Consequences of Urban Development in Alexandria, Egypt." *Environment and Urbanization*, Volume 2, No. 1, April.

Hardoy, Jorge A. and David Satterthwaite. 1992. *Environmental Problems in Third World Cities: An Agenda for the Poor and the Planet*. London: International Institute for Environment and Development.

Hardoy, Jorge A. and David Satterthwaite. 1989. *Squatter Citizen: Life in the Urban Third World*. London: Earthscan: Earthscan Publications Limited.

Hardoy, Jorge A. and David Satterthwaite. 1984. "Third World Cities and the Environment of Poverty." *Geoforum*, Volume 15, No. 3.

Harpham et al., eds. 1988. *In the Shadow of the City: Community Health and the Urban Poor*. Oxford: Oxford University Press.

Hasan, Arif. 1992. Personal communication.

Hauchman, Fred S. 1991. "Air Pollution and Health in Developing Countries." Prepared for U.S Agency for International Development Environmental Health Workshop, Washington, D.C., February 14-15.

Hufschmidt, Maynard et al. 1987. "Water Management Policy Options for the Beijing-Tianjin Region of China." Report by Environment and Policy Institute, East West Center, Honolulu, Hawaii.

International Institute for Environment and Development. 1992. "Sustainable Cities: Meeting Needs, Reducing Resource Use and Recycling, Re-use and Reclamation." *Environment and Urbanization*, Volume 4, No. 2, October.

Japan International Cooperation Agency. 1990. "Seminar on Medium to Long-Term Improvement/ Management Plan of Road and Road Transport in Bangkok," January.

Khalifa, Ahmed M. and Mohamed M. Moheieoldin in *The Metropolis Era*. 1983. Vol. 1. Eds. Mattei Dogon and John C. Casada. Sage Publications.

Kaji, Gautam. 1992. "East Asia: Economic Miracles, Environmental Problems." *The Bank's World*, March.

Karmokolias, Yannis. 1990. "Automotive Industry Trends and Prospects for Investment in Developing Countries." International Finance Corporation Discussion Paper No. 7. Washington, D.C.: World Bank.

Khan, Akhter Hameed. 1992. "Orangi Pilot Project Programs." Karachi, Pakistan: Orangi Pilot Project-Research and Training Institute. 2ed, December.

Kreimer, Alcira and Edward G. Echeverria.1991. "Case Study: Housing Reconstruction in Mexico City." In *Managing Natural Disasters and the Environment*. Eds. Alcira Kreimer and Mohan Munasinghe. Washington, D.C.: World Bank.

Kreimer, Alcira and Mohan Munasinghe. 1991. "Managing Environmental Degradation and Natural Disasters: An Overview. In *Managing Natural Disasters and the Environment*. Eds. Alcira Kreimer and Mohan Munasinghe. Washington, D.C.: World Bank.

Kreimer, Alcira et al. 1992. "Towards a Sustainable Urban Environment: The Rio de Janeiro Study." World Bank Draft Report.

Kulaba, Saitiel. 1989. "Local Government and the Management of Urban Services in Tanzania." In *African Cities in Crisis*, Eds. Richard E. Stren and Rodney R. White. Boulder: Westview Press.

Leitmann, Josef. 1991. *Energy-Environment Linkages in the Urban Sector*. Urban Management Programme Discussion Paper No. 2. Washington, D.C.: World Bank.

———. 1991. "Environmental Profile of São Paulo." Draft Urban Management Programme Case Study, August.

———. 1991. "Environmental Profile of Tianjin." Draft Urban Management Programme Case Study, August.

______. 1994. *Rapid Urban Environmental Assessment: Lessons from Cities of the Developing World.* Urban Management Programme Discussion Papers Nos. 14–15. Washington, D.C.: World Bank.

Leonard, H. Jeffrey. 1986. "Hazardous Wastes: The Crisis Spreads." *Asian National Development,* April.

Levinson, Arik and Sudhir Shetty. 1992. "Efficient Environmental Regulation: Case Studies of Urban Air Pollution Los Angeles, Mexico City, Cubatao, and Ankara." World Bank Policy Research Working Paper No. 942, Washington, D.C.

Lowe, Marcia D. 1992. "City Limits." *World Watch,* Volume 5, No. 1, February.

Margulis, Sergio. 1992. "Back of the Envelope Estimates of Environmental Damage Costs in Mexico." World Bank Policy Research Working Paper 824, Washington, D.C.

Mayo, S.K. and S. Angel. 1993. "Housing: Enabling Markets to Work," World Bank Policy Paper, Washington, D.C.

Megalli, Nabil. 1991. "Turning Industrial Relations Green." *Our Planet,* Volume 3, No. 6.

Moguel, Julio and E. Velazquez. 1992. "Urban Social Organization and Ecological Struggle in Durango, Mexico." In *Grassroots Environmental Action: People's Participation in Sustainable Development.* Eds. D. Ghai and J. Vivian. London: Routledge.

Motor Columbus. 1985. "Metropolitan Area of Douala, Study of Waste Management and Resource Recovery." Part A, Phase I. December.

Munasinghe, Mohan, Braz Menezes, and Martha Preece. 1991. "Case Study: Rio Flood Reconstruction and Prevention Project." In *Managing Natural Disasters and the Environment.* Eds. Alcira Kreimer and Mohan Munasinghe. Washington, D.C.: World Bank.

National Institute of Urban Affairs. 1988. "State of India's Urbanization." New Delhi.

———. 1990. "Sulabh Shauchalaya: An Innovative Response to Urban Sanitation Needs." Mega-Cities Case Study, New Delhi.

New York Times. 1992. "At Least 129 Dead as Blasts Rock a Mexican City." April 23.

______. 1992. "Nine Officials of Oil Trust and City to Face Charges in Mexico Blasts." April 27.

______. 1992. "Test Was Halted Before Explosion." April 24.

Organization Economic Cooperation and Development. 1992. "Cities and the Environment." Draft report. Paris, September.

_______. 1993. "Energy and Land Use Planning." Paris.

Paulsson, Bengt. 1992. *Satellite Remote Sensing and GIS Analysis for Urban Applications.* Urban Management Programme Discussion Paper, No. 9. Washington, D.C.: World Bank.

Pean, Leslie. 1993. "AGETIP: A New Resource to Meet the Urban Challenge." *Infrastructure Notes*, Urban No. OU-8. World Bank Transportation, Water, and Urban Development Department. Washington, D.C., February.

Pearce, David W. and Jeremy Warford. Forthcoming. *World Without End: Economics, Environment, and Sustainable Development.* New York: Oxford University Press.

Rabinovitch, Jonas with Josef Leitmann. 1993. *Environmental Innovation and Management in Curitiba, Brazil.* Urban Management Programme Working Paper No. 1, June. Washington, D.C.: World Bank.

Ribeiro, Edgar F. 1987. "Improved Sanitation and Environmental Health Conditions: An Evaluation of Sulabh International's Low Cost Sanitation Project in Bihar." 3d ed. New Delhi: Sulabh Interational Patna.

Sahabat Alam Malaysia. 1983. The State of MalaysianEnvironment 1983–84—Towards Greater Environmental Awareness. Penang, Malaysia: Friends of the Earth.

Schaefer, Morris. 1991. "Combatting Environmental Pollution: National Capabilities for Health Protection." Geneva: World Health Organization.

Schteingart, Martha. 1989. "The Environmental Problems Associated with Urban Development in Mexico City." *Environment and Urbanization.* Vol. 1, No. 1, April.

Serageldin, Ismail. 1992. "Historic Cairo: A Proposal for Public-Private Partnership." Paper presented to the Urban Sector Workshop, Egyptian Ministry of Development, New Communities, Housing, and Public Utilities, and the World Bank. Cairo, July.

Shin, Eusoon, Regina Gregory, Maynard Hufschmidt, Yok-Shiu Lee, James E. Nickum, and Chieko Umitsu. 1991. "Economic Valuation of the Urban Environment With Emphasis on Asia and Non-Productivity Approaches." Draft paper prepared for the Urban Management Programme, October.

Shrivastava, Paul. 1987. *Bhopal: Anatomy of a Crisis*. Cambridge, Massachusetts: Ballinger Publishing Company.

Sivaramakrishnan, K.C. 1992. "India's Urban Environment Status Report." Calcutta: The Times Research Foundation, September.

Sivaramakrishnan, K.C. and Leslie Green. 1986. *Metropolitan Management: The Asian Experience*. New York: Oxford University Press.

Smith, Kirk. 1993. "Fuel Combustion, Air Pollution Exposure, and Health: The Situation in Developing Countries." *Annual Review of Energy and Environment,* Number 18: 529-566.

Thomann, Robert J. 1972. "Comprehensive Water Quality Management—the Delaware Estuary." In *Systems Analysis and Water Quality Management.* New York: McGraw-Hill.

United Nations Centre for Human Settlements and United Nations Environment Programme. 1987. "Environmental Guidelines for Settlements Planning and Management (3 Vols)." Nairobi.

United Nations Development Programme. 1992. *The Urban Environment in Developing Countries.* New York.

U.S. Agency for International Development. 1992. "Success Stories of Urban Environmental Management." Case studies prepared for USAID (RHUDO/Asia).

U.S. Agency for International Development and U.S. Environmental Protection Agency. 1990. "Ranking Environmental Health Risks in Bangkok, Thailand." December.

Vaughn, J.P. 1990. "Lessons from Experience." In *Health Systems Decentralization*. Ed. Mills, A. et al. Geneva: World Health Organization.

Water and Sanitation for Health Project (WASH). 1993. *Voices from the City,* Volume No. 3, Arlington, Virginia, September.

Washington Post. 1993. "Garbage Slide Kills 22 Salvadorans." June 17.

Watson, Peter. 1992. "Urban Forum Presentation on the Agency for the Execution of Works in the Public Interest (AGETIP)," April 22.

Wheeler, David. 1992. "The Economics of Industrial Pollution Control: An International Perspective." Industry and Energy Department Working Paper, Industry Series Paper No. 60. Washington, D.C.: World Bank.

Wilson, Leslie. 1993. "A Dumpsite Guilty of the Greatest Crime." *Wastes Management* 83 (10).

World Bank Infrastructure and Urban Development Department. 1991. "Urban Policy and Economic Development: An Agenda for the 1990s." Washington, D.C.

World Bank Operations Evaluation Department. 1992. "Annual Review of Evaluation Results 1991." Report No. 11062, August.

World Bank. 1993. "Water Resources Management." World Bank Policy Paper, Washington, D.C.

______. 1990. *World Development Report 1990: Poverty*. New York: Oxford University Press.

______. 1992. *World Development Report 1992: Development and the Environment*. New York: Oxford University Press.

———. 1993. *World Development Report 1993: Investing in Health.* New York: Oxford University Press.

World Health Organization. 1989. New Approaches to Improve Road Safety. Technical Report Series 781. Geneva.

———. 1988. *Urbanization and Its Implications for Child Health: Potential for Action.* Geneva.

______. 1992. *The International Drinking Water Supply and Sanitation Decade: Review of Decade Progress (as of December 1990).* Geneva.

______. 1992. *Our Planet, Our Health.* Geneva.

Distributors of World Bank Publications

ARGENTINA
Carlos Hirsch, SRL
Galeria Guemes
Florida 165, 4th Floor-Ofc. 453/465
1333 Buenos Aires

AUSTRALIA, PAPUA NEW GUINEA, FIJI, SOLOMON ISLANDS, VANUATU, AND WESTERN SAMOA
D.A. Information Services
648 Whitehorse Road
Mitcham 3132
Victoria

AUSTRIA
Gerold and Co.
Graben 31
A-1011 Wien

BANGLADESH
Micro Industries Development Assistance Society (MIDAS)
House 5, Road 16
Dhanmondi R/Area
Dhaka 1209

Branch offices:
Pine View, 1st Floor
100 Agrabad Commercial Area
Chittagong 4100

BELGIUM
Jean De Lannoy
Av. du Roi 202
1060 Brussels

CANADA
Le Diffuseur
151A Boul. de Mortagne
Boucherville, Québec
J4B 5E6

Renouf Publishing Co.
1294 Algoma Road
Ottawa, Ontario
K1B 3W8

CHILE
Invertec IGT S.A.
Av. Santa Maria 6400
Edificio INTEC, Of. 201
Santiago

CHINA
China Financial & Economic Publishing House
8, Da Fo Si Dong Jie
Beijing

COLOMBIA
Infoenlace Ltda.
Apartado Aereo 34270
Bogota D.E.

COTE D'IVOIRE
Centre d'Edition et de Diffusion Africaines (CEDA)
04 B.P. 541
Abidjan 04 Plateau

CYPRUS
Center of Applied Research
Cyprus College
6, Diogenes Street, Engomi
P.O. Box 2006
Nicosia

DENMARK
SamfundsLitteratur
Rosenoerns Allé 11
DK-1970 Frederiksberg C

DOMINICAN REPUBLIC
Editora Taller, C. por A.
Restauración e Isabel la Católica 309
Apartado de Correos 2190 Z-1
Santo Domingo

EGYPT, ARAB REPUBLIC OF
Al Ahram
Al Galaa Street
Cairo

The Middle East Observer
41, Sherif Street
Cairo

FINLAND
Akateeminen Kirjakauppa
P.O. Box 128
SF-00101 Helsinki 10

FRANCE
World Bank Publications
66, avenue d'Iéna
75116 Paris

GERMANY
UNO-Verlag
Poppelsdorfer Allee 55
D-5300 Bonn 1

HONG KONG, MACAO
Asia 2000 Ltd.
46-48 Wyndham Street
Winning Centre
2nd Floor
Central Hong Kong

HUNGARY
Foundation for Market Economy
Dombovari Ut 17-19
H-1117 Budapest

INDIA
Allied Publishers Private Ltd.
751 Mount Road
Madras - 600 002

Branch offices:
15 J.N. Heredia Marg
Ballard Estate
Bombay - 400 038

13/14 Asaf Ali Road
New Delhi - 110 002

17 Chittaranjan Avenue
Calcutta - 700 072

Jayadeva Hostel Building
5th Main Road, Gandhinagar
Bangalore - 560 009

3-5-1129 Kachiguda Cross Road
Hyderabad - 500 027

Prarthana Flats, 2nd Floor
Near Thakore Baug, Navrangpura
Ahmedabad - 380 009

Patiala House
16-A Ashok Marg
Lucknow - 226 001

Central Bazaar Road
60 Bajaj Nagar
Nagpur 440 010

INDONESIA
Pt. Indira Limited
Jalan Borobudur 20
P.O. Box 181
Jakarta 10320

IRAN
Kowkab Publishers
P.O. Box 19575-511
Tehran

IRELAND
Government Supplies Agency
4-5 Harcourt Road
Dublin 2

ISRAEL
Yozmot Literature Ltd.
P.O. Box 56055
Tel Aviv 61560

ITALY
Licosa Commissionaria Sansoni SPA
Via Duca Di Calabria, 1/1
Casella Postale 552
50125 Firenze

JAPAN
Eastern Book Service
Hongo 3-Chome, Bunkyo-ku 113
Tokyo

KENYA
Africa Book Service (E.A.) Ltd.
Quaran House, Mfangano Street
P.O. Box 45245
Nairobi

KOREA, REPUBLIC OF
Pan Korea Book Corporation
P.O. Box 101, Kwangwhamun
Seoul

Korean Stock Book Centre
P.O. Box 34
Yeoeido
Seoul

MALAYSIA
University of Malaya Cooperative Bookshop, Limited
P.O. Box 1127, Jalan Pantai Baru
59700 Kuala Lumpur

MEXICO
INFOTEC
Apartado Postal 22-860
14060 Tlalpan, Mexico D.F.

NETHERLANDS
De Lindeboom/InOr-Publikaties
P.O. Box 202
7480 AE Haaksbergen

NEW ZEALAND
EBSCO NZ Ltd.
Private Mail Bag 99914
New Market
Auckland

NIGERIA
University Press Limited
Three Crowns Building Jericho
Private Mail Bag 5095
Ibadan

NORWAY
Narvesen Information Center
Book Department
P.O. Box 6125 Etterstad
N-0602 Oslo 6

PAKISTAN
Mirza Book Agency
65, Shahrah-e-Quaid-e-Azam
P.O. Box No. 729
Lahore 54000

PERU
Editorial Desarrollo SA
Apartado 3824
Lima 1

PHILIPPINES
International Book Center
Suite 1703, Cityland 10
Condominium Tower 1
Ayala Avenue, H.V. dela Costa Extension
Makati, Metro Manila

POLAND
International Publishing Service
Ul. Piekna 31/37
00-677 Warzawa

For subscription orders:
IPS Journals
Ul. Okrezna 3
02-916 Warszawa

PORTUGAL
Livraria Portugal
Rua Do Carmo 70-74
1200 Lisbon

SAUDI ARABIA, QATAR
Jarir Book Store
P.O. Box 3196
Riyadh 11471

SINGAPORE, TAIWAN, MYANMAR, BRUNEI
Gower Asia Pacific Pte Ltd.
Golden Wheel Building
41, Kallang Pudding, #04-03
Singapore 1334

SOUTH AFRICA, BOTSWANA
For single titles:
Oxford University Press Southern Africa
P.O. Box 1141
Cape Town 8000

For subscription orders:
International Subscription Service
P.O. Box 41095
Craighall
Johannesburg 2024

SPAIN
Mundi-Prensa Libros, S.A.
Castello 37
28001 Madrid

Librería Internacional AEDOS
Consell de Cent, 391
08009 Barcelona

SRI LANKA AND THE MALDIVES
Lake House Bookshop
P.O. Box 244
100, Sir Chittampalam A. Gardiner Mawatha
Colombo 2

SWEDEN
For single titles:
Fritzes Fackboksforetaget
Regeringsgatan 12, Box 16356
S-103 27 Stockholm

For subscription orders:
Wennergren-Williams AB
P. O. Box 1305
S-171 25 Solna

SWITZERLAND
For single titles:
Librairie Payot
Case postale 3212
CH 1002 Lausanne

For subscription orders:
Librairie Payot
Service des Abonnements
Case postale 3312
CH 1002 Lausanne

THAILAND
Central Department Store
306 Silom Road
Bangkok

TRINIDAD & TOBAGO, ANTIGUA BARBUDA, BARBADOS, DOMINICA, GRENADA, GUYANA, JAMAICA, MONTSERRAT, ST. KITTS & NEVIS, ST. LUCIA, ST. VINCENT & GRENADINES
Systematics Studies Unit
#9 Watts Street
Curepe
Trinidad, West Indies

UNITED KINGDOM
Microinfo Ltd.
P.O. Box 3
Alton, Hampshire GU34 2PG
England